TRAINING FOR CERTIFI

Taking An ASE Certification Test

This study guide will help prepare you to take and pass the ASE test. It contains descriptions of the types of questions used on the test, the task list from which the test questions are derived, a review of the task list subject information, and a practice test containing ASE style questions.

ABOUT ASE

The National Institute for Automotive Service Excellence (ASE) is a non-profit organization founded in 1972 for the purpose of improving the quality of automotive service and repair through the voluntary testing and certification of automotive technicians. Currently, there are over 400,000 professional technicians certified by ASE in over 40 different specialist areas.

ASE certification recognizes your knowledge and experience, and since it is voluntary, taking and passing an ASE certification test also demonstrates to employers and customers your commitment to your profession. It can mean better compensation and increased employment opportunities as well.

ASE not only certifies technician competency, it also promotes the benefits of technician certification to the motoring public. Repair shops that employ at least one ASE technician can display the ASE sign. Establishments where 75% of technicians are certified, with at least one technician certified in each area of service offered by the business, are eligible for the ASE Blue Seal of Excellence program. ASE encourages consumers to patronize these shops through media campaigns and car care clinics.

To become ASE certified, you must pass at least one ASE exam and have at least two years of related work experience. Technicians that pass all tests in a series earn Master Technician status. Your certification is valid for five years, after which time you must retest to retain certification, demonstrating that you have kept up with the changing technology in the field.

THE ASE TEST

An ASE test consists of forty to eighty multiple-choice questions. Test questions are written by a panel of technical experts from vehicle, parts and equipment manufacturers, as well as working technicians and technical education instructors. All questions have been pre-tested and quality checked on a national sample of technicians. The questions are derived from information presented in the task list, which details the knowledge that a technician must have to pass an ASE test and be recognized as competent in that category. The task list is periodically updated by ASE in response to changes in vehicle technology and repair techniques.

© Advanstar 2003.7

Customer Service 1-800-240-1968
FAX 218-723-9437
e-mail www.ma@superfill.com
URL: www.motorage.com

Taking An ASE Certification Test

There are five types of questions on an ASE test:

Direct, or Completion
MOST Likely
Technician A and Technician B
EXCEPT
LEAST Likely

Direct, or Completion

This type of question is the kind that is most familiar to anyone who has taken a multiple-choice test: you must answer a direct question or complete a statement with the correct answer. There are four choices given as potential answers, but only one is correct. Sometimes the correct answer to one of these questions is clear, however in other cases more than one answer may seem to be correct. In that case, read the question carefully and choose the answer that is most correct. Here is an example of this type of test question:

A compression test shows that one cylinder is too low. A leakage test on that cylinder shows that there is excessive leakage. During the test, air could be heard coming from the tail pipe. Which of the following could be the cause?

    A. broken piston rings
    B. bad head gasket
    C. bad exhaust gasket
    D. an exhaust valve not seating

There is only one correct answer to this question, answer **D**. If an exhaust valve is not seated, air will leak from the combustion chamber by way of the valve out to the tail pipe and make an audible sound. Answer C is wrong because an exhaust gasket has nothing to do with combustion chamber sealing. Answers A and B are wrong because broken rings or a bad head gasket would have air leaking through the oil filler or coolant system.

MOST Likely

This type of question is similar to a direct question but it can be more challenging because all or some of the answers may be nearly correct. However, only one answer is the most correct. For example:

When a cylinder head with an overhead camshaft is discovered to be warped, which of the following is the most correct repair option?

    A. replace the head
    B. check for cracks, straighten the head, surface the head
    C. surface the head, then straighten it
    D. straighten the head, surface the head, check for cracks

The most correct answer is **B**. It makes no sense to perform repairs on a cylinder head that might not be useable. The head should first be checked for warpage and cracks. Therefore, answer B is more correct than answer D. The head could certainly be replaced, but the cost factor may be prohibitive and availability may be limited, so answer B is more correct than answer A. If the top of the head is warped enough to interfere with cam bore alignment and/or restrict free movement of the camshaft, the head must be straightened before it is resurfaced, so answer C is wrong.

Technician A and Technician B

These questions are the kind most commonly associated with the ASE test. With these questions you are asked to choose which technician statement is correct, or whether they both are correct or incorrect. This type of question can be difficult because very often you may find one technician's statement to be clearly correct or incorrect while the other may not be so obvious. Do you choose one technician or both? The key to answering these questions is to carefully examine each technician's statement independently and judge it on its own merit. Here is an example of this type of question.

A vehicle equipped with rack-and-pinion steering is having the front end inspected. Technician A says that the inner tie-rods-ends should be inspected while in their normal running position. Technician B says that if movement is felt between the tie-rod stud and the socket while the tire is moved in and out, the inner tie-rod should be replaced. Who is correct?

    A. Technician A
    B. Technician B
    C. Both A and B
    D. Neither A or B

The correct answer is **C**; both technicians' statements are correct. Technician B is clearly correct because any play felt between the tie-rod stud and the socket while the tire is moved in and out indicates that the assembly is worn and requires replacement. However, Technician A is also correct because inner tie-rods should be inspected while in their normal running position, to prevent binding that may occur when the suspension is allowed to hang free.

EXCEPT

This kind of question is sometimes called a negative question because you are asked to give the incorrect answer. All of the possible answers given are correct EXCEPT one. In effect, the correct answer to the question is the one that is wrong. The word EXCEPT is always capitalized in these questions. For example:

All of the following are true of torsion bars **EXCEPT:**
  A. They can be mounted longitudinally or transversely.
  B. They serve the same function as coil springs.
  C. They are interchangeable from side-to-side.
  D. They can be used to adjust vehicle ride height.

The correct answer is **C.** Torsion bars are not normally interchangeable from side-to-side. This is because the direction of the twisting or torsion is not the same on the left and right sides. All of the other answers contain true statements regarding torsion bars.

LEAST Likely

This type of question is similar to EXCEPT in that once again you are asked to give the answer that is wrong. For example:

Blue-gray smoke comes from the exhaust of a vehicle during deceleration. Of the following, which cause is **LEAST** likely?
  A. worn valve guides
  B. broken valve seals
  C. worn piston rings
  D. clogged oil return passages

The correct answer is **C.** Worn piston rings will usually make an engine smoke worse under acceleration. All of the other causes can allow oil to be drawn through the valve guides under the high intake vacuum that occurs during deceleration.

PREPARING FOR THE ASE TEST

Begin preparing for the test by reading the task list. The task list describes the actual work performed by a technician in a particular specialty area. Each question on an ASE test is derived from a task or set of tasks in the list. Familiarizing yourself with the task list will help you to concentrate on the areas where you need to study.

The text section of this study guide contains information pertaining to each of the tasks in the task list. Reviewing this information will prepare you to take the practice test.

Take the practice test and compare your answers with the correct answer explanations. If you get an answer wrong and don't understand why, go back and read the information pertaining to that question in the text.

After reviewing the tasks and the subject information, and taking the practice test, you should be prepared to take the ASE test or be aware of areas where further study is needed. When studying with this study guide or any other source of information, use the following guidelines to make sure the time spent is as productive as possible:

• Concentrate on the subject areas where you are weakest.
• Arrange your schedule to allow specific times for studying.
• Study in an area where you will not be distracted.
• Don't try to study after a full meal or when you are tired.
• Don't wait until the last minute and try to 'cram' for the test.

TAKING THE ASE TEST

Make sure you get a good night's sleep the night before the test. Have a good lunch on test day but either eat lightly or skip dinner until after the test. A heavy meal will make you tired.

Bring your admission ticket, some form of photo identification, three or four sharpened #2 pencils and a watch (to keep track of time as the test room may not have a clock) with you to the test center.

The test proctor will explain how to fill out the answer sheet and how much time is allotted for each test. You may take up to four certification tests in one sitting, but this may prove too difficult unless you are very familiar with the subject areas.

When the test begins, open the test booklet to see how many questions are on the test. This will help you keep track of your progress against the time remaining. Mark your answer sheet clearly, making sure the answer number and question number correspond.

Read through each question carefully. If you don't know the answer to a question and need to think about it, move on to the next question. Don't spend too much time on any one question. After you have worked through to the end of the test, check your remaining time and go back and answer the questions you had trouble with. Very often, infor-

3

Taking An ASE Certification Test

mation found in questions later in the test can help answer some of the ones with which you had difficulty.

If you are running out of time and still have unanswered test questions, guess the answers if necessary to make sure every question is answered. Do not leave any answers blank. It is to your advantage to answer every question, because your test score is based on the number of correct answers. A guessed answer could be correct, but a blank answer can never be.

To learn exactly where and when the ASE Certification Tests are available in your area, as well as the costs involved in becoming ASE certified, please contact ASE directly for a registration booklet.

The National Institute for Automotive Service Excellence 101 Blue Seal Drive, S.E. Suite 101 Leesburg, VA 20175

1-877-ASE-TECH (273-8324)

http://www.asecert.org

TABLE OF CONTENTS

Test Specifications And Task List 6

Steering Systems Diagnosis And Repair 10

Suspension Systems Diagnosis And Repair 34

Related Suspension And Steering Service 54

Wheel Alignment Diagnosis, Adjustment And Repair 62

Wheel And Tire Diagnosis And Repair 72

Sample Test Questions 82

Answers To Sample Test Questions 92

Glossary 98

Suspension And Steering

TEST SPECIFICATIONS
FOR SUSPENSION AND STEERING (TEST A4)

CONTENT AREA		NUMBER OF QUESTIONS IN ASE TEST	PERCENTAGE OF COVERAGE IN ASE TEST
A. Steering System Diagnosis And Repair		10	25%
1. Steering Columns	(3)		
2. Steering Units	(4)		
3. Steering Linkage	(3)		
B. Suspension Systems Diagnosis And Repair		11	28%
1. Front Suspensions	(6)		
2. Rear Suspensions	(5)		
C. Related Suspension And Steering Service		2	5%
D. Wheel Alignment Diagnosis, Adjustment And Repair		12	30%
E. Wheel And Tire Diagnosis And Repair		5	13%
	Total	40	100%

The five-year Recertification Test will cover the same content areas as those listed above. However, the number of questions in each content area of the Recertification Test will be reduced by about one-half.

The following pages list the tasks covered in each content area. These task descriptions offer detailed information to technicians preparing for the test, and to persons who may be instructing technicians in Suspension And Steering. The task list may also serve as a guideline for question writers, reviewers and test assemblers.

It should be noted that the number of questions in each content area might not be equivalent to the number of tasks listed. Some tasks are complex and broad in scope, and may be covered in several questions. Other tasks are simple and narrow in scope; one question may cover many tasks. The main purpose for listing the tasks is to accurately describe what skills are required for certification in this area, not to make each task correspond to a particular test question.

SUSPENSION AND STEERING TEST TASK LIST

A. STEERING SYSTEMS DIAGNOSIS AND REPAIR
(10 questions)

1. Steering Columns
(3 questions)

Task 1 - Diagnose steering column noises and steering effort concerns (including manual and electronic tilt and telescoping mechanisms); determine needed repairs.

Task 2 - Inspect and replace steering column, steering shaft U-joint(s), flexible coupling(s), collapsible columns, steering wheels (includes steering wheels with air bags and/or other steering wheel mounted controls and components).

Task 3 - Disarm, enable and properly handle airbag system components during vehicle service following manufacturers' procedures.

2. Steering Units
(4 questions)

Task 1 - Diagnose steering gear (non-rack-and-pinion type) noises, binding, vibration, free-play, steering effort, steering pull (lead) and leakage concerns; determine needed repairs.

Task 2 - Diagnose rack-and-pinion steering gear noises, binding, vibration, free-play, steering effort, steering pull (lead) and leakage concerns; determine needed repairs.

Task 3 - Inspect power steering fluid level and condition; adjust level in accordance with vehicle manufacturers' recommendations.

Task 4 - Inspect, adjust, align and replace power steering pump belt(s) and tensioners.

Task 5 - Diagnose power steering pump noises, vibration and fluid leakage; determine needed repairs.

Task 6 - Remove and replace power steering pump; inspect pump mounting and attaching brackets; remove and replace power steering pump and pulley.

Task 7 - Inspect and replace power steering pump seals, gaskets, reservoir and valves.

Task 8 - Perform power steering system pressure and flow tests; determine needed repairs.

Task 9 - Inspect and replace power steering hoses, fittings, O-rings, coolers and filters.

Task 10 - Remove and replace steering gear (non-rack-and-pinion type).

Task 11 - Remove and replace rack-and-pinion steering gear; inspect and replace mounting bushings and brackets.

Task 12 - Adjust steering gear (non-rack-and-pinion type) worm bearing preload and sector lash.

Task 13 - Inspect and replace steering gear (non-rack-and-pinion type) seals and gaskets.

Task 14 - Adjust rack-and-pinion steering gear.

Task 15 - Inspect and replace rack-and-pinion steering gear bellows boots.

Task 16 - Flush, fill and bleed power steering system.

Task 17 - Diagnose, inspect, repair or replace components of variable-assist steering systems.

3. Steering Linkage
(3 questions)

Task 1 - Inspect and adjust (where applicable) front and rear steering linkage geometry (including parallelism and vehicle ride height).

Task 2 - Inspect and replace pitman arm.

Task 3 - Inspect and replace center link (relay rod/drag link/intermediate rod).

Task 4 - Inspect, adjust (where applicable) and replace idler arm(s) and mountings.

Task 5 - Inspect, replace and adjust tie-rods, tie-rod sleeves/adjusters, clamps and tie-rod ends (sockets/bushings).

Task 6 - Inspect and replace steering linkage damper(s).

B. SUSPENSION SYSTEMS DIAGNOSIS AND REPAIR
(13 questions)

1. Front Suspensions
(6 questions)

Task 1 - Diagnose front suspension system noises, body sway/roll and ride height concerns; determine needed repairs.

Task 2 - Inspect and replace upper and lower control arms, bushings and shafts

Task 3 - Inspect and replace rebound and jounce bumpers.

Task 4 - Inspect, adjust and replace strut rods/radius arms (compression/tension) and bushings.

Task 5 - Inspect and replace upper and lower ball joints (with or without wear indicators).

Task 6 - Inspect non-independent front axle assembly for bending, warpage and misalignment.

Task 7 - Inspect and replace front steering knuckle/spindle assemblies and steering arms.

Task 8 - Inspect and replace front suspension system coil springs and spring insulators (silencers).

Task 9 - Inspect and replace front suspension system leaf spring(s), leaf spring insulators (silencers), shackles, brackets, bushings and mounts.

Task 10 - Inspect, replace and adjust front suspension system torsion bars and mounts.

Task 11 - Inspect and replace front stabilizer bar (sway bar) bushings, brackets and links.

Task 12 - Inspect and replace front strut cartridge or assembly.

Task 13 - Inspect and replace front strut bearing and mount.

2. Rear Suspensions
(5 questions)

Task 1 - Diagnose rear suspension system noises, body sway/roll and ride height concerns; determine needed repairs.

Task 2 - Inspect and replace rear suspension system coil springs and spring insulators (silencers).

Task 3 - Inspect and replace rear suspension system lateral links/arms (track bars), control (trailing) arms, stabilizer bars (sway bars),

bushings and mounts.

Task 4 - Inspect and replace rear suspension system leaf spring(s), leaf spring insulators (silencers), shackles, brackets, bushings and mounts.

Task 5 - Inspect and replace rear rebound and jounce bumpers.

Task 6 - Inspect and replace rear strut cartridge or assembly and upper mount assembly.

Task 7 - Inspect non-independent rear axle assembly for bending, warpage and misalignment.

Task 8 - Inspect and replace rear ball joints and tie-rod/toe link assemblies.

Task 9 - Inspect and replace rear knuckle/spindle assembly.

C. RELATED SUSPENSION AND STEERING SERVICE
(2 questions)

Task 1 - Inspect and replace shock absorbers, mounts and bushings.

Task 2 - Diagnose and service front and/or rear wheel bearings.

Task 3 - Diagnose, inspect, adjust, repair or replace components (including sensors, switches and actuators) of electronically controlled suspension systems (including primary and supplemental air suspension and ride control systems).

Task 4 - Inspect and repair front and/or rear cradle (crossmember/subframe) mountings, bushings, brackets and bolts.

Task 5 - Diagnose, inspect, adjust, repair or replace components (including sensors, switches and actuators) of electronically controlled steering systems.

Task 6 - Diagnose, inspect, repair or replace components of power steering idle speed compensation systems.

D. WHEEL ALIGNMENT DIAGNOSIS, ADJUSTMENT AND REPAIR
(12 questions)

Task 1 - Diagnose vehicle wander, drift, pull, hard steering, bump steer (toe curve), memory steer, torque steer and steering return concerns; determine needed repairs.

Task 2 - Measure vehicle ride

7

height; determine needed repairs.

Task 3 - Measure front and rear wheel camber; determine needed repairs.

Task 4 - Adjust front and/or rear wheel camber on suspension systems with a camber adjustment.

Task 5 - Measure caster; determine needed repairs.

Task 6 - Adjust caster on suspension systems with a caster adjustment.

Task 7 - Measure and adjust front wheel toe.

Task 8 - Center steering wheel.

Task 9 - Measure toe-out on turns (turning radius/angle); determine needed repairs.

Task 10 - Measure SAI/KPI (steering axis inclination/kingpin inclination); determine needed repairs.

Task 11 - Measure included angle; determine needed repairs.

Task 12 - Measure rear wheel toe; determine needed repairs or adjustments.

Task 13 - Measure thrust angle; determine needed repairs or adjustments.

Task 14 - Measure front wheelbase setback/offset; determine needed repairs or adjustments.

Task 15 - Check front cradle (crossmember/subframe) alignment; determine needed repairs or adjustments.

D. WHEEL AND TIRE DIAGNOSIS AND REPAIR
(5 questions)

Task 1 - Diagnose tire wear patterns; determine needed repairs.

Task 2 - Inspect tire condition, size and application (load and speed ratings).

Task 3 - Measure and adjust tire air pressure.

Task 4 - Diagnose wheel/tire vibration, shimmy and noise concerns; determine needed repairs.

Task 5 - Rotate tires/wheels and torque fasteners according to manufacturers' recommendations.

Task 6 - Measure wheel, tire, axle flange and hub runout (radial and lateral); determine needed repairs.

Task 7 - Diagnose tire pull (lead) problems; determine corrective actions.

Task 8 - Dismount and mount tire on wheel.

Task 9 - Balance wheel and tire assembly.

The preceding Task List data details all of the related informational subject matter you are expected to know in order to sit for this ASE Certification Test. Your own years of experience as a technician in the professional automotive service repair trade also should provide you with added background.

Finally, a conscientious review of the self-study material provided in this Training for ASE Certification unit also should help you to be adequately prepared to take this test.

NOTES

TRAINING FOR CERTIFICATION

STEERING SYSTEMS DIAGNOSIS AND REPAIR

The steering system allows the driver to turn the vehicle in a safe and predictable manner. The basic steering system consists of the steering wheel and column, the steering gear, and the linkage that connects the gear to the steering knuckles. The two most common system designs in use today are the conventional worm gear recirculating ball steering system and the rack-and-pinion steering system.

In the conventional system, the worm gear is housed in a steering box and connected to the steering column shaft. When the shaft turns, the worm gear causes the ball nut, which rides on roller balls, to move up and down the worm gear. The ball nut turns a sector gear and shaft that is connected to the pitman arm. The pitman arm transfers the gearbox motion to the steering linkage. Various steering linkage designs are used with a conventional gearbox, however the most common is the parallelogram, which uses a center link, idler arm and tie-rod assemblies to connect to the steering knuckles.

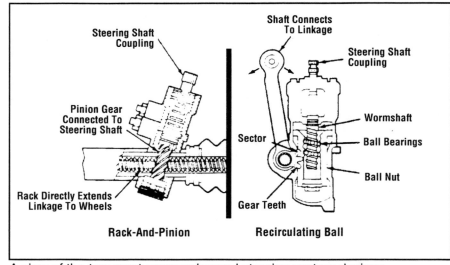

A view of the two most commonly used steering system designs. (Courtesy: DaimlerChrysler Corp.)

In the rack-and-pinion system, the steering shaft is connected to a pinion gear in the rack and pinion housing. When the shaft turns, the pinion gear acts on the rack gear causing the rack to slide sideways in the housing. The ends of the rack are connected to the tie-rod assemblies, which are the only steering linkage used in the rack-and-pinion system.

Power steering systems use a crankshaft-driven belt to turn a hydraulic pump to provide boost for certain difficult steering operations, particularly during parking when the wheel may be stationary on the ground.

The power steering pump is usually a vane-and-impeller type,

We employ technicians certified by the National Institute for
AUTOMOTIVE SERVICE EXCELLENCE
Let us show you their credentials

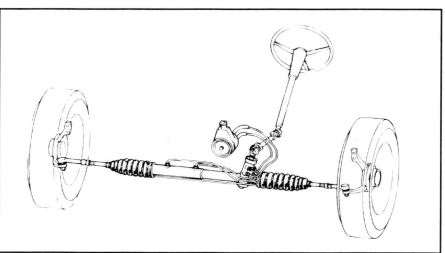

A power assisted rack-and-pinion steering gear.

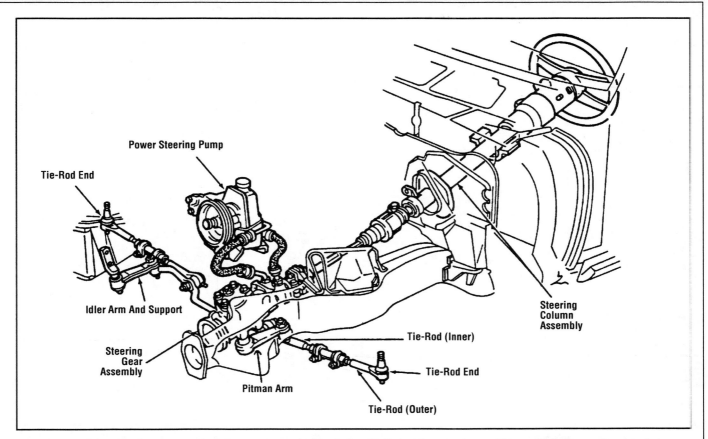

Here is an example of a conventional worm gear recirculating ball steering system with parallelogram steering linkage.

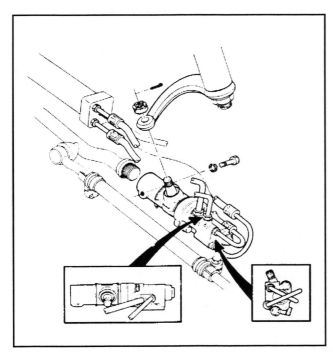

Control valve and power cylinder of a non-integral power steering system.
(Courtesy: GM Corp.)

fitted internally with a pressure-relief valve that opens when the hydraulic boost reaches its maximum. Typically, this happens only when the steering wheel is turned all the way to the stop and held there. You can hear the opening of the pressure relief valve. The pump pressurizes hydraulic fluid and delivers it through one or more high-pressure hoses to the steering gear. The oil then forces the rack or the gear to the side the driver requires, with the pressure metered through a valve responsive to the degree of effort the driver uses to turn the steering wheel.

There are two power steering system designs, integral and non-integral. In the non-integral design, the pressure and return hoses from the pump are connected to a control valve located on the end of the center link and attached to the pitman arm. Two hoses from the control valve are connected to a power cylinder, which is also attached to the center link, with the power cylinder ram mounted to a frame bracket. The non-integral system was used on older vehicles.

The integral system is the most common power steering system in use today. In the integral system, the power cylinder and control valve are contained in one housing, either in a conventional steering gear or a rack-and-pinion assembly.

11

TRAINING FOR CERTIFICATION

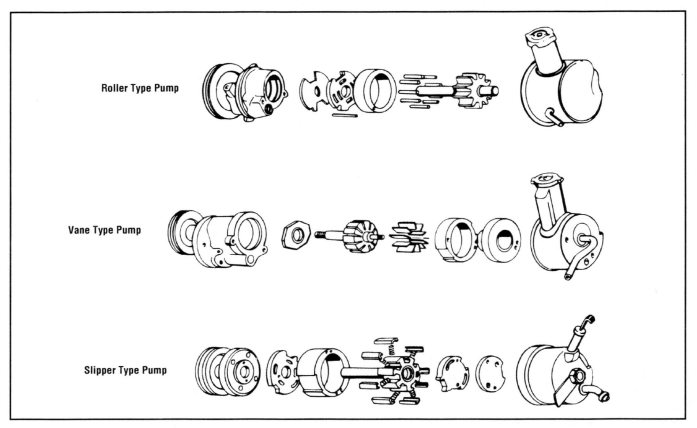

There are three power steering pump designs: Roller, Vane and Slipper. The most commonly used design is the vane type pump. Pumps can have the fluid reservoir incorporated in the pump assembly as shown, or the reservoir can be mounted remotely. *(Courtesy: Federal Mogul/Moog Automotive Division)*

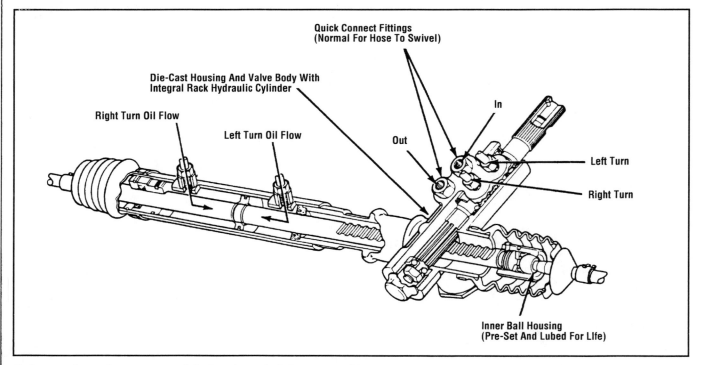

Cutaway view of a power assisted rack-and-pinion assembly.

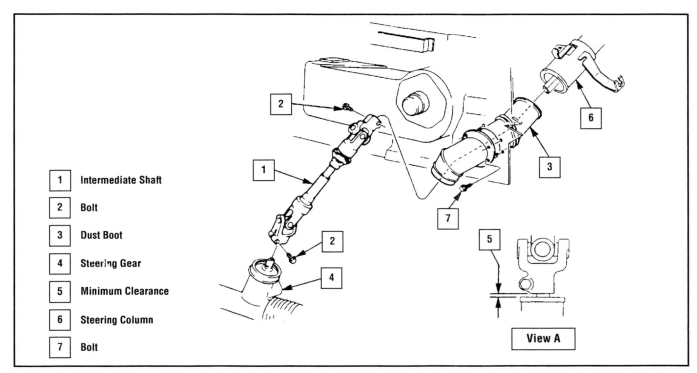

On some vehicles, the steering column is connected to the steering gear with an intermediate shaft.
(Courtesy GM Corp.)

STEERING COLUMNS

The driver turns the steering gear via the steering wheel, which is mounted on the steering column. The typical steering column contains the steering shaft and shaft support bearings, as well as brackets that attach the column to the firewall and the dash panel. The steering wheel is splined to the steering shaft and secured with a nut.

Most steering columns also include small U-joints or flexible couplings that connect the steering shaft to the steering gear. Many vehicles, especially 4WD light trucks, use an intermediate shaft which runs from the lower end of the steering column shaft to the steering gear input shaft where it connects with a flex coupling or U-joint. The upper end of the intermediate shaft is equipped with a U-joint which is, usually, not replaceable or rebuildable.

On vehicles with a driver's side air bag, the SIR (Supplemental Inflatable Restraint or, air bag) system must be disarmed and the air bag removed before the steering wheel or steering column is removed. Failure to do so may cause accidental deployment of the air bag, and may result in personal injury.

Steering Column Diagnosis

Check for looseness in the steering column shaft bearings and loose or broken steering column mounts by moving the steering wheel up-and-down and from right-to-left. Turn the steering wheel from stop-to-stop and check for looseness or binding. If the vehicle is equipped with power steering, this should be done with the engine running so that the power steering system is functional.

Excessive steering wheel play can be caused by worn steering column U-joints and/or steering column bearings. Hard steering or poor steering return can be caused by binding steering col-

The SIR (Supplemental Inflatable Restraint) system must be disabled before removing the air bag inflator module and steering wheel.
(Courtesy: Federal Mogul/Moog Automotive Division)

TRAINING FOR CERTIFICATION

umn U-joints. Normally, a worn or damaged U-joint can be easily replaced. Most flexible couplings are replaceable or rebuildable, with kits available, but some of the flexible couplings are fixed parts of the shaft and can't be replaced separately.

Many steering columns have telescoping and tilting mechanisms. Check the telescoping and tilting mechanisms for full range of motion. The most common problem will be that the steering wheel won't move to a desired position, or that the telescoping/tilting function does not work at all. These systems are not complicated, but there are too many of them to consider here. Check the appropriate service manual for the repair procedures.

All vehicles made over the last few decades' feature collapsible steering columns to prevent the column from heavily impacting the driver during a front-end collision. Manufacturers use several mechanisms to facilitate the steering wheel moving toward the dash if the driver's torso hits it. This can be done using U-joints offset, sliding tubes, or deformed metal, which bends on impact.

If there has been an accident and the steering column has partially collapsed, even if the steering still works, there is no repair possible; the column must be replaced. To determine whether a column has collapsed, begin by visually inspecting the steering wheel and steering column cover. There should be a gap between the two. If this is closed, the column is almost certainly collapsed. Push and pull on the steering wheel to check for play. Inspect the column where it leaves the firewall for out of position parts. Remove the steering column cover and look for movement of any 'sliding' parts or collapse of accordion sections. Manufacturers publish length measurements to check for column collapse. This can only be done with the column removed from the vehicle.

Steering Column Service

AIR BAG SYSTEM

Before servicing the steering wheel or steering column, the air bag must be properly disarmed. In general, this usually involves disconnecting the negative battery cable from the battery terminal and taping the cable end to prevent it from accidentally contacting the battery terminal. Then, wait at least 10 minutes for the backup power supply to discharge. Always consult the vehicle service manual for the exact disarming procedure.

When working on the air bag system, wear eye protection and follow all safety precautions. After an air bag module has been removed, carry the module with the cover pad facing away from

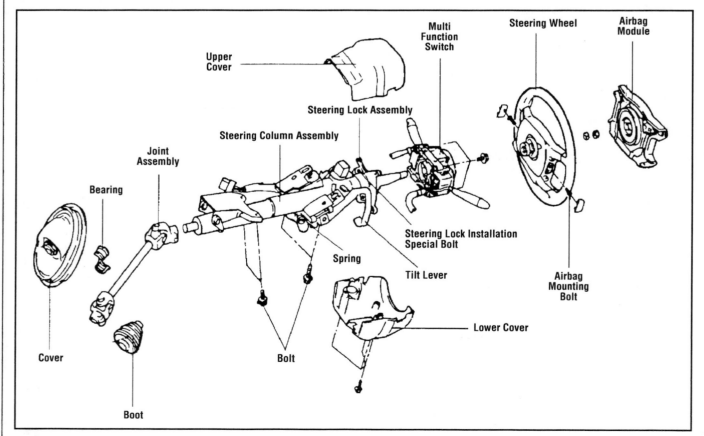

Exploded view of a typical steering column assembly. *(Courtesy: Hyundai Motor Co.)*

14

the body. Store the module with the cover pad facing up, so that accidental deployment does not launch it into the air.

STEERING COLUMN REMOVAL AND INSTALLATION

To remove the steering column, first position the vehicle's front wheels in the straight-ahead position and make sure the column is in the LOCK position. If equipped, properly disable the air bag according to the manufacturer's instructions.

Remove the inflator module or horn pad from the steering wheel and disconnect all wiring between the steering wheel and column. Remove the steering wheel retaining nut from the steering shaft and then check for alignment marks on the steering wheel and shaft. If none are present, scribe marks so the wheel can be reinstalled in the same position. Remove the steering wheel from the column using a steering wheel puller.

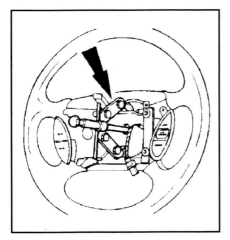

Removing the steering wheel using a puller.
(Courtesy: Ford Motor Co.)

CAUTION: Never hammer on the end of the steering shaft or use a knock off type puller in an attempt to remove the steering wheel, as damage to the shaft, column or bearings may occur.

Remove the necessary column and dash trim panels to gain access to the steering column wiring and mounting bolts. Disconnect the column wiring and shifter/transmission range indicator linkage.

Scribe alignment marks on the steering shaft and intermediate shaft or steering gear so the column can be reinstalled in the same position. Remove the U-joint fasteners or flexible coupling pinch bolt to disconnect the steering shaft from the intermediate shaft or steering gear. Support the column and remove the column mounting bolts. Remove the column from the vehicle.

Installation of the column is the reverse of removal. Be sure to refer to the vehicle service manual for torque specifications.

STEERING UNITS

Steering System Diagnosis

The most common steering system complaints are excessive free-play or looseness, hard steering, pulling, vibration and noise. These can be caused by incorrect adjustment, component wear, lack of lubrication or incorrect wheel alignment.

Begin your diagnosis by road testing the vehicle to verify the complaint. Then, visually inspect the steering system, starting with the tires. Make sure the tires are the correct size and are properly inflated. Check the tread for unusual wear patterns. Mismatched, underinflated, worn or unbalanced tires can cause pulling, wander, vibration and hard steering.

With the wheels in the straight-ahead position, check the steering wheel free-play. Turn the steering wheel back and forth and see how far the steering wheel can be moved before the front wheels begin to move. As a rule of thumb, if you can move the steering wheel more than 1 1/2-in. before moving the front wheels, there is excessive free-play. However, be sure to check the service manual for the exact specification for the vehicle in question.

Excessive free-play can be caused by steering gear that is out of adjustment or by worn steering system components. Begin your inspection at the steering wheel and work out towards the wheels. Check for looseness in the steering column shaft bearings and loose or broken steering column mounts by moving the steering wheel up-and-down and from right-to-left. Have an assistant turn the steering wheel back-and-forth while you check for worn or noisy steering shaft U-joints and couplings.

On vehicles with conventional steering, first check the steering gear mounting bolts, then grasp the pitman arm and attempt to move the sector shaft back-and-forth in the steering gear housing. Check the idler arm bushings and all the steering gear linkage ball studs and sockets for looseness. Make sure the threaded portions of the tie-rod ends are not loose in the tie-rod sleeves. If all components pass visual inspection, the excessive free-play is most likely caused by a loose steering gear worm bearing preload adjustment.

On vehicles with rack-and-pinion steering, check the steering gear mounting bolts for looseness and check the mounting bushings for wear and damage. Grasp the pinion gear shaft and attempt to move it in-and-out of the gear; if there is movement, the pinion bearing preload requires adjustment. Squeeze the steering rack bellows until the inner tie-rod socket can be felt, then push and pull on the wheel. If there is looseness in the socket, the inner tie-rod is worn. Make this inspection with the tie-rod assemblies in their normal running position. If the inner tie-rods are inspected with the suspension and steering linkage hanging in the full rebound position, the inner tie-rod may rest against the socket housing and hide any looseness. In some cases, the boot may need to be removed in order to inspect the inner tie-rod. Also check the ball stud and socket at the outer

TRAINING FOR CERTIFICATION

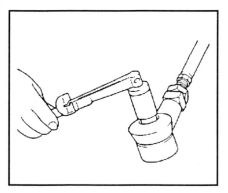

Measuring tie-rod ball stud turning effort with a torque wrench. *(Courtesy: Hyundai Motor Co.)*

tie-rod for looseness.

Hard steering can be caused by binding or damaged steering system components, lack of lubrication and incorrect steering gear adjustment. Check the steering shaft U-joints and the lower shaft seal for binding and the dash panel seal for interference and listen for binding or scraping noises. If the vehicle subframe was recently lowered during service and there is binding in the steering column, the subframe and column may have been incorrectly repositioned during reassembly. Binding may occur in a rack-and-pinion system if the rack gear is damaged. Insufficient lubricant in the steering gear and steering linkage can also cause binding. Make sure the steering gear is filled with the proper type and quantity of lubricant.

Check the steering linkage for binding by rotating the tie-rod ends. Disconnect the tie-rod ends from the steering linkage and lock two nuts on the stud threads. Measure the force required to turn the stud in its socket with a torque wrench and compare the measurement with manufacturer's specifications.

Problems in the power assist portion of the steering system include fluid leaks, noise, loss of power assist, increased steering effort and hard steering. These problems can be caused by low fluid level, a loose drive belt, air in the system or components that are worn or in need of adjustment.

The most common problem with power steering is power steering fluid leakage. Leaks are a safety hazard for two reasons, both of them serious.

First, a power steering leak will mean loss of steering assist boost, and, to an inexperienced driver, a change in the feel of the steering wheel may seem like lockup. After an accident, motorists who have had a powering steering boost loss often describe the situation as something like, "The steering wheel froze in place," when all that really happened was loss of the power assist.

The second hazard is fire. While power steering fluid is not easily flammable, it will burn when pressurized and sprayed in atomized form on a hot exhaust manifold. The possibility of an oil-fueled engine compartment fire at vehicle speed is not something to ignore.

Check the hoses, hose connections, power steering gear, pump and reservoir for signs of fluid leakage. If the source of the leak

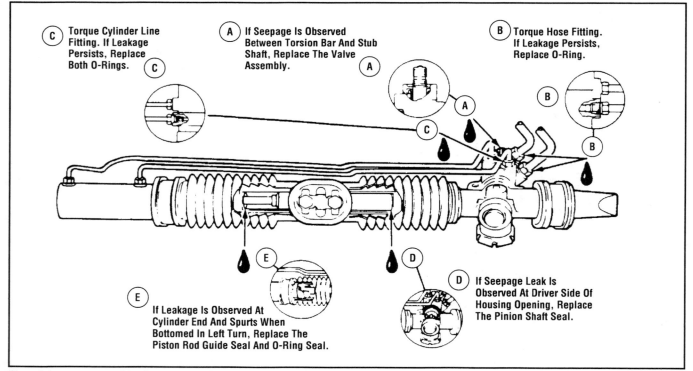

Possible sources of leaks on a typical power rack-and-pinion steering gear. *(Courtesy: GM Corp.)*

is not obvious, clean the area, then run the engine, turning the steering wheel from stop-to-stop to pressurize the system. Look for the presence of fresh fluid.

The next most common power steering complaint is noise, most often from belt squeal. Inspect the power steering pump drive belt tension and look for signs of cracking, glazing or chunking on the belt, which would warrant replacement. A loose or worn belt can cause erratic or high steering effort. In the most extreme case, a cracked or worn belt could break, causing complete loss of steering assist.

Another type of noise is whining or growling coming from the power steering pump, caused by low fluid level or air in the system. A low fluid level could cause erratic or increased steering effort, as well as cause possible damage to the pump and steering gear. A hissing noise, most often heard when the wheels are turned and the vehicle is not moving, is caused by normal relief valve operation and is not an indicator of a power steering system problem.

Steering effort complaints can also be caused by problems in the pump and/or steering gear. Steering effort can be measured using a spring scale. Let the engine idle for several minutes and turn the steering wheel from stop-to-stop several times to warm the power steering fluid. With the engine running and the wheels on a clean dry floor, attach the spring scale to the steering wheel and measure the pull required to turn the steering wheel. Compare the measurement with manufacturer's specifications.

Excessive steering effort can be caused by a lack of power assist or by binding of mechanical components. Lack of power assist in only one direction can be caused by a leaking wormshaft oil seal ring or a broken ring on the worm piston. Lack of power assist in both directions can be caused by a broken ring on the

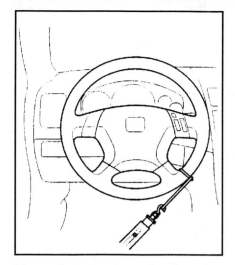
Measuring steering wheel turning effort with a spring scale. *(Courtesy: Honda Motor Co.)*

power piston or a bad power steering pump. Performing a system pressure test can help identify the source of problems.

Steering System Service

POWER STEERING FLUID LEVEL INSPECTION

Power steering fluid reservoirs can either be integral with the power steering pump or remotely mounted. Remote reservoirs are usually made of transparent plastic and allow the fluid level to be quickly checked against the graduations on the side of the

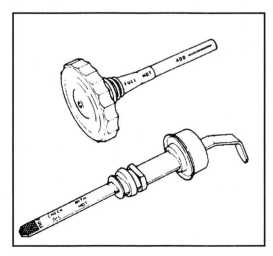
Typical power steering pump integral reservoir dipsticks.

reservoir. The fluid level in integral reservoirs is usually checked with a dipstick. Both reservoirs and dipsticks usually have FULL COLD and FULL HOT or MIN and MAX level indicators.

It is usually best to check fluid level with the fluid at operating temperature, however, make sure the fluid level is at least at the FULL COLD or MIN level before operating the engine to prevent pump or steering gear damage. If the level is checked with a dipstick, before removing it wipe the cap and the area around the reservoir opening with a clean cloth to prevent dirt from entering the system.

Operate the vehicle to bring the fluid to operating temperature. Park the vehicle on level ground and set the parking brake. Turn the engine OFF.

If equipped with a transparent reservoir, check the fluid level against the graduations on the side of the reservoir. If equipped with a dipstick, remove the dipstick from the reservoir and wipe it with a clean cloth. Reinsert the dipstick fully into the reservoir, remove it and check the level. Add or remove fluid as necessary.

When checking the fluid level, also note its condition. Discolored fluid or evidence of particles in the fluid may be indicative of wear or damage to the pump, steering gear or hoses. Fluid that is foamy is caused by air in the system.

POWER STEERING PUMP BELT
Adjustment

Most vehicles are equipped with serpentine V-ribbed belts that are kept in adjustment with automatic tensioners. The older V-belt designs must be periodically manually adjusted.

Belt tension can be checked using the deflection method or by using a belt tension gauge. Locate a point midway between the longest accessible belt span. If using the deflection

17

TRAINING FOR CERTIFICATION

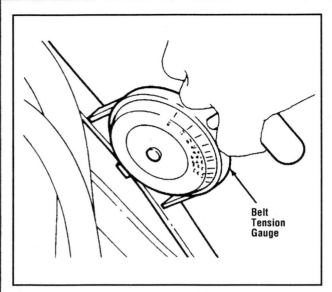

Checking belt tension using a belt tension gauge. (Courtesy: Ford Motor Co.)

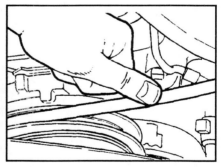

Checking belt tension using the deflection method.

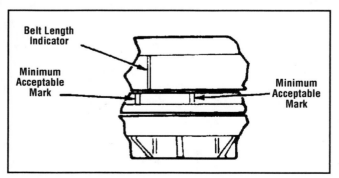

Belt length indicator on an automatic belt tensioner. (Courtesy: Ford Motor Co.)

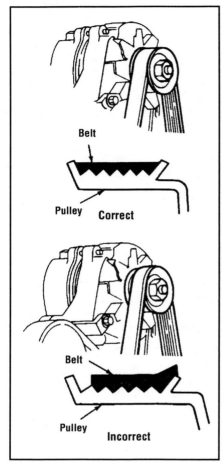

Make sure V-ribbed belts are properly seated in the pulley grooves. One revolution of the engine with the belt incorrectly seated can damage the belt.
(Courtesy: Ford Motor Co.)

method, push on the belt with your finger using moderate pressure and measure the belt deflection. If you are using a belt tension gauge, position the gauge and measure the amount of force necessary to deflect the belt. Compare your reading with specifications.

The belt tension should be checked on vehicles with automatic belt tensioners to make sure the tensioner is functioning properly. Some automatic tensioners are equipped with belt length indicator and minimum and maximum acceptable marks, the theory being that if the correct length belt is installed on the engine and the mark is within range, belt tension is correct.

To adjust V-belt tension, loosen the adjuster pulley, or pump pivot and adjuster bolts. Then use a suitable prybar to move the pulley or pump bracket until the belt tension is correct. Never pry directly against the power steering pump housing as it can be damaged. Tighten all fasteners and recheck belt tension.

Some adjuster pulleys and pumps are equipped with a threaded shaft for adjustment. With this design, loosen the locknut and turn the shaft with a wrench or socket until the correct belt tension is obtained. Tighten the locknut and recheck belt tension.

Removal And Installation

Removal and installation of V-belts usually involves loosening the adjuster pulley or pump pivot and adjuster bolts, moving the pulley or pump to eliminate belt tension and removing the belt. It may be necessary to remove other accessory drive belts to gain access to the power steering pump belt. Install the new belt and adjust belt tension.

Before removing a serpentine V-ribbed belt, make sure there is a belt routing diagram handy or draw one prior to belt removal, to prevent installation problems. Use a socket or wrench to tilt the automatic tensioner away from the belt, and then remove the belt from the pulleys. Inspect the pulleys after the belt is removed and look for damage to the pulleys and wear or dirt in the pulley grooves that could damage the new belt. Clean or replace pulleys

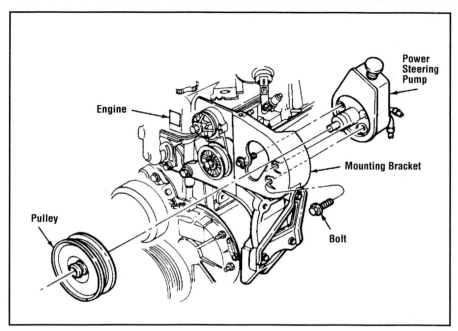

On this vehicle, the pump pulley must be removed before the pump can be removed from the vehicle.
(Courtesy: Ford Motor Co.)

as necessary. When installing the belt, make sure it is routed properly and that it is properly seated in the pulley grooves.

POWER STEERING PUMP
Removal And Installation

To remove the power steering pump, first remove the pump drive belt. Clean any dirt away from the hose connections to prevent dirt from entering the system. Disconnect the remote reservoir connections, if equipped, and the pressure and return hoses from the pump. Cap or plug the hoses and pump fittings.

Support the pump, remove the pump mounting bolts and remove the pump from the vehicle. If the pump is being replaced, remove the pump pulley and install it on the new pump. On some vehicles, the pulley must be removed before the pump is removed from the mounting bracket.

Place the pump in position and install the mounting bolts. Tighten the bolts or leave them loose at this time if they are also used to set belt tension. Connect the hoses to the pump. Install the pump drive belt and adjust the tension. Fill the pump reservoir with fluid and bleed the system. Road test the vehicle and check for leaks.

Overhaul

If a power steering pump is worn or leaking, it is usually replaced, however, a leaking pump can be resealed. Due to the many design variations, a comprehensive and detailed discussion of power steering pump rebuilding is beyond the scope of this study guide. The following describes the overhaul of a typi-

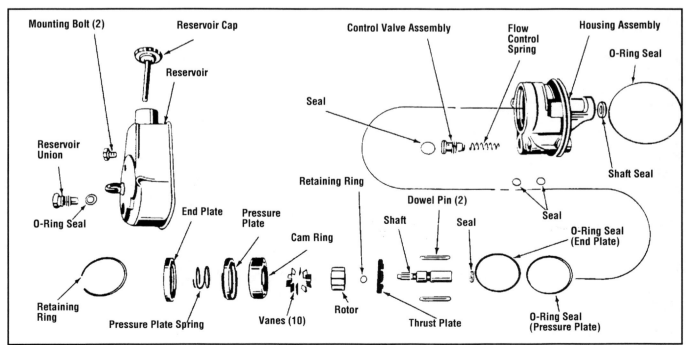

Disassembled view of a typical power steering pump. *(Courtesy: Ford Motor Co.)*

19

TRAINING FOR CERTIFICATION

cal power steering pump. Always consult the proper service manual for the specific disassembly and assembly procedure for the pump you are servicing.

Remove the power steering pump from the vehicle and remove the pulley. Inspect the pump driveshaft bearing or bushing for wear. In some cases the bearing or bushing cannot be serviced separately, requiring pump replacement if it is worn. Position the front hub of the pump in a vise with the shaft pointing down. Do not clamp the pump too tightly so as not to damage the driveshaft bearing.

Remove the pressure line fitting, O-ring and reservoir mounting bolts from the back of the reservoir. Twist the reservoir back-and-forth slightly to unseat the reservoir O-ring seal, and then remove the reservoir from the pump housing. Remove the pressure fitting and reservoir mounting bolt seals from the pump housing.

Pry the end plate retaining ring from the housing, and then remove the end plate and pressure plate spring. Remove the pump housing from the vise and turn it over to allow the control valve and spring to fall out. Remove the end plate O-ring, and then tap lightly on the end of the shaft until the pressure plate falls free.

Remove the cam ring and vanes, and then push the end of the shaft through the housing to remove the shaft and rotor assembly. Remove the pressure plate and end plate O-ring seals from the housing bore and remove the shaft seal.

Clean all parts with solvent and dry with compressed air. Inspect all parts for wear and damage. Replace the pump if necessary.

Install a new shaft seal using a seal installer. Lubricate a new pressure plate O-ring seal with power steering fluid and install it in the housing bore. Lubricate the driveshaft and install it in the housing.

Install the cam ring and vanes. Lubricate the pressure plate with power steering fluid and install it in the housing. Lubricate the end plate O-ring seal and install it in the housing bore. Lubricate the end plate, and then press it into position enough to allow installation of the retaining ring.

Lubricate the control valve with power steering fluid and install it in the pump housing with the spring. Install a new reservoir mounting bolt and the pressure fitting seals on the housing. Lubricate a new reservoir O-ring seal and install it on the housing. Lubricate the edge of the reservoir and install it on the housing with the mounting bolts. Lubricate a new O-ring and install the pressure fitting.

Install the power steering pump pulley and install the pump in the vehicle.

Pulley Replacement

Inspect the pulley for damage and misalignment. A bent or damaged pulley can shorten belt life, damage pump shaft bearings and seals and cause vibration. A misaligned pulley can also shorten belt life, cause noise and in severe cases cause the belt to jump from the pulley.

Pulley removal is required for replacement, but in many cases it is also necessary in order to remove the pump from its mounting bracket. Never use a hammer to attempt to remove the pulley. Various pullers are available to remove pulleys; some have jaws that grasp the pulley from behind and some from a groove at the front of the pulley. All rotate the forcing screw against the pump driveshaft.

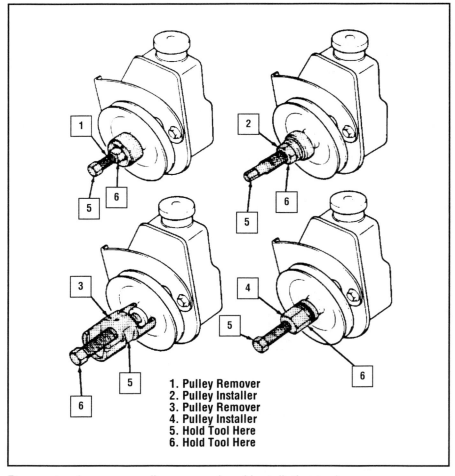

Power steering pump pulley removal and installation tools. *(Courtesy: GM Corp.)*

1. Pulley Remover
2. Pulley Installer
3. Pulley Remover
4. Pulley Installer
5. Hold Tool Here
6. Hold Tool Here

Pulley installation tools have bolts that thread into the end of the pump driveshaft. The bolt is then held steady while the installer presses the pulley into place. Check pulley alignment as the pulley is installed; many pulleys are not installed until they bottom on a shoulder but rather just until they align.

POWER STEERING SYSTEM PRESSURE TESTING

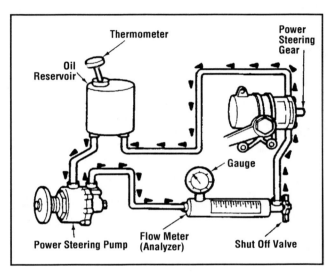

Power steering system pressure testing. *(Courtesy: Ford Motor Co.)*

Increased steering effort, erratic steering and lack of power assist can all be caused by problems with the power steering pump or gear. Performing a power steering system pressure test can determine the cause of the problem. A suitable pressure gauge designed for this purpose along with pressure specifications for the vehicle in question are required to perform the test. The pressure gauge is connected on the pressure side between the power steering pump and gear, with the shutoff valve toward the steering gear.

Before beginning the test, make sure the power steering fluid reservoir is filled to the proper level. Place a thermometer in the reservoir fluid, then run the engine at idle while turning the steering wheel from stop-to-stop to warm the fluid.

When the fluid temperature reaches about 170°F, check and record the pressure with the wheels in the straight-ahead position. If the pressure is above specification, usually about 150 psi, check for restricted hoses or a damaged steering gear.

WARNING: During the next part of the test you must completely close the shutoff valve. Do not close the valve for more than 5 seconds at one time, as excessive pump pressure could cause a power steering hose to rupture, possibly causing personal injury.

Completely close and partially open the shutoff valve three times, noting the highest pressure each time the valve is closed. Compare your pressure readings with specifications. If each of the readings is within specification and there is less than 50 psi difference between readings, the power steering pump is functioning properly. If the readings are within range but there is more than 50 psi difference between readings, clean or replace the flow control valve or replace the pump. If the pressure readings are below specification, replace the flow control valve and recheck. If the pressures are still low, replace the pump.

Turn the steering wheel from stop-to-stop; however do not keep the steering wheel at full stop for more than a few seconds. Record the highest pressure reading at full stop and compare with the highest pump pressure recorded in the previous step. If the pressure is the same, the steering gear is functioning properly. If the pressure is lower, the steering gear is leaking internally and must be repaired or replaced.

POWER STEERING HOSES

Power steering hoses are designed to withstand the high pressure and temperature generated in the power steering system. When replacing a power steering hose, always use the one designed for the specific function and application. Inspect hoses for leaks, cracks, swelling or other damage and replace as required.

Use a flare nut wrench to loosen and tighten power steering hose fittings. The pressure hose will have a metal fitting at each end. The return, or low-pressure hose will be connected to the steering gear with a fitting but may sometimes be secured to the pump with a clamp. Some systems use O-rings at the fittings, which should be replaced when a hose is installed.

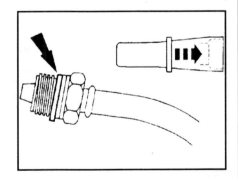

Always use new sealing O-rings when connecting a fitting. A tool like this can be used when installing a new O-ring to prevent damaging the seal on the fitting threads.
(Courtesy: Ford Motor Co.)

Route the new hose in the same manner as the original, and so it will not rub against other components. Make sure the hose is not twisted; most hoses have ridged surfaces or a stripe running their length to provide a visual reference. Reinstall any clips or ties that may be used to keep the hoses from fouling on other components.

TRAINING FOR CERTIFICATION

After hose installation, bleed the system and check for leaks.

CONVENTIONAL STEERING GEAR

Adjustment is usually the only service required on the steering gearbox, however, if the box is worn to the point where adjustment is insufficient, the gearbox must be rebuilt or replaced.

Manual Steering Gear Adjustment

Worm gear bearing preload is the compression force on the bearings directly turned by the steering column. If there is too little preload, the steering will feel loose. If there is too much preload the steering will feel stiff. To measure the preload, remove the steering linkage from the system so only the internal loads of the gearbox remain. Ordinarily, it is better to remove the center link from the pitman arm than to remove the pitman arm from the sector shaft, though both techniques would have the same unloading effect.

With the linkage removed, measure the resistance of the steering wheel to turning. This can be done either with a very light torque wrench, calibrated in inch lbs., or with a spring scale attached to the steering wheel. Consult the vehicle service manual for the specified method.

A worn steering gearbox will typically have too little worm bearing preload rather than too much. There are two kinds of adjustment: by shims or by an adjuster nut or plug. If the system uses shims, removing/adding a shim will increase/decrease the preload, once you retorque the cover bolts to specifications. If the system uses an adjuster, loosen the locknut and tighten the adjuster nut or plug to increase preload or loosen it to decrease preload.

Steering gear lash or steering gear worm/sector mesh preload determines how much free-play there is in the steering wheel. The free-play referred to here is only what comes from the gearbox itself, not any additional looseness caused by looseness in the rest of the linkage. Again, disconnect the steering linkage, then turn the wheel to either extreme, counting the number of turns and keeping track of the angles at the stop (this will also tell you whether the steering wheel is centered in the gearbox range). Sector shaft mesh will be tightest just at the center of travel. With use and wear, the gearbox mesh will tend to wear loose. Measure the steering wheel's resistance to travel right at the middle of its range, using either a low-reading torque wrench or the aforementioned spring scale method. Adjust the mesh by loosening the locknut on the sector shaft adjusting screw, turning the screw in until you reach the correct lash, and then holding the screw positioned with a screwdriver while you retighten the locknut.

Sector shaft axial play can only be adjusted (when at all) with the gearbox disassembled. Check for specific instructions in the shop manual.

Power Steering Gear Adjustment

Adjustments to conventional

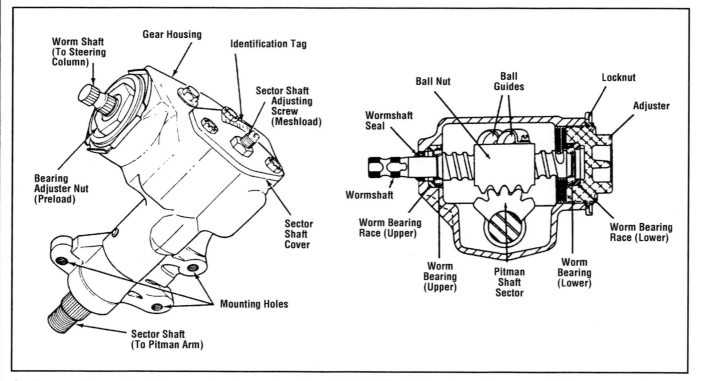

On the gearbox on the left, worm bearing preload is adjusted with the bearing adjuster nut. On the one on the right, worm bearing preload is adjusted with the adjuster plug.

power steering gears are generally the same as those detailed for conventional manual steering gears. Before adjusting mesh preload, the manufacturer may specify that the fluid be drained from the steering gear by disconnecting the return hose, placing the end in a container and turning the steering wheel from stop-to-stop several times.

In addition to worm bearing preload and worm/sector mesh preload, the power steering control valve can be checked for centering and sometimes adjusted. If the control valve is not centered, the vehicle may drift, pull or have poor returnability.

Always refer to the vehicle service manual for the specific adjustment instructions.

Removal And Installation

If adjustment does not correct a gearbox problem and overhaul or replacement is necessary, remove the gearbox as follows:

Position the front wheels in the straight-ahead position. If the vehicle is equipped with an airbag, center the steering wheel and secure it in that position. It is extremely important that this step is followed prior to disconnecting the steering column from the gearbox; if the steering wheel moves while disconnecting or reconnecting the steering column, the clock spring will be damaged.

Remove the pinch bolt retaining the steering shaft coupling to the gearbox. Scribe alignment marks on the steering shaft and gearbox worm gear shaft so they can be reassembled in the same position. If equipped, disconnect and plug the power steering pressure and return hoses.

Raise and safely support the vehicle. Remove the nut from the sector shaft, and then use a suitable puller to remove the pitman arm from the sector shaft. Most pitman arms are indexed to the shaft, but if not, scribe alignment marks prior to removal so the pitman arm can be reinstalled in the same position.

Support the gearbox and

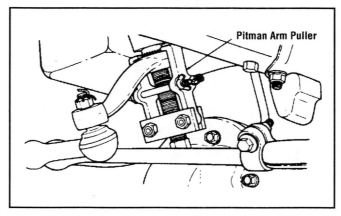
Removing the pitman arm from the sector shaft. *(Courtesy: Ford Motor Co.)*

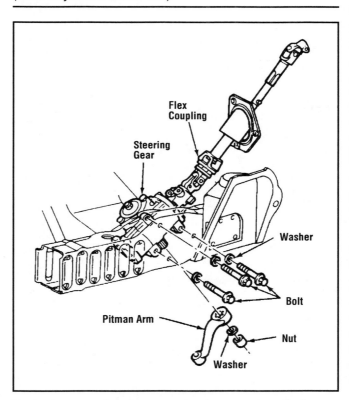
Typical conventional power steering gear installation. *(Courtesy: Ford Motor Co.)*

remove the gearbox retaining bolts. Separate the gearbox from the steering shaft and remove the gearbox from the vehicle.

Before reinstalling the steering gear, turn the worm shaft from stop-to-stop, counting the number of turns. Turn the worm shaft half the number of turns from stop to center the steering gear. Align the worm gear shaft with the steering shaft and index the pitman arm with the sector shaft or align the marks made at removal. Torque all fasteners to specification. If equipped with power steering, bleed the system and check for leaks.

Power Steering Gear Resealing

A leaking steering gear is usually replaced, however, it can be resealed. Due to the many design variations, a comprehensive and detailed discussion of steering gear rebuilding is beyond the scope of this study guide. The following describes the disassembly and assembly of a typical power steering gear. Always consult the proper service manual for the specific disassembly and assembly procedure for the gear you are servicing.

Remove the steering gear from the vehicle and drain the fluid into a suitable container. Thoroughly clean the outside of the gear, and then mount the steering gear in a vise.

Center the worm gear shaft. Remove the sector shaft cover bolts, then tap on the lower end of the sector shaft with a soft mallet to loosen the cover. Remove the cover and sector shaft as an assembly and discard the O-ring.

TRAINING FOR CERTIFICATION

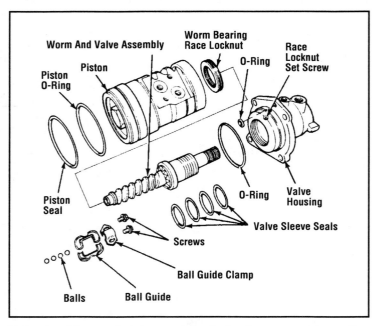

Valve housing and piston and control valve assembly portion of a power steering gearbox. *(Courtesy: Ford Motor Co.)*

Remove the valve housing bolts and remove the valve housing and piston and control valve assembly from the steering gear housing. Remove the ball guide clamps, and then position the piston and valve assembly over a clean container. Let the ball guides fall into the container, and then turn the wormshaft from stop-to-stop until all balls fall into the container. Remove the wormshaft and valve assembly from the piston. Remove the worm bearing locknut and remove the wormshaft and valve assembly from the valve housing. Remove and discard the valve seals.

Remove the wormshaft seal from the valve housing cover and the sector shaft seal from the steering gear housing.

Clean and inspect components for wear and damage. Replace parts or the entire steering gear, as necessary.

Install new wormshaft and sector shaft seals. Lubricate new valve seals with power steering fluid and install them on the wormshaft and valve assembly. Install the wormshaft and valve assembly into the valve housing. Install the worm bearing locknut and torque to specification.

Insert the wormshaft and the valve assembly into the piston. Place the ball guides in the piston and install the balls. Rotate the wormshaft from stop-to-stop, if necessary, to install all the balls. Install the ball guide clamps.

Install the valve housing and piston and control valve assembly into the steering gear housing, using new O-rings. Rotate the piston so it will mesh properly with the sector shaft. Tighten the valve housing bolts to specification.

Install a new sector shaft cover O-ring in the steering gear housing. Turn the wormshaft to center the piston. Apply petroleum jelly to the sector shaft journal and install the sector shaft and cover in the steering gear housing. Tighten the bolts to specification.

Adjust the gear mesh preload. Install the steering gear in the vehicle. Bleed the power steering system. Road test and check for leaks and proper steering operation.

RACK-AND-PINION STEERING GEAR

On-vehicle service of rack and pinion steering gear is usually limited to tie-rod assembly repair. Depending on the design, adjustment may or may not require that the unit be removed from the vehicle. Always consult the vehicle service manual to be sure.

Removal And Installation

Position the front wheels in the straight-ahead position. If the vehicle is equipped with an airbag, center the steering wheel and secure it in that position. It is extremely important that this step is followed prior to disconnecting the steering column from the pinion gear shaft; if the steering wheel moves while disconnecting or reconnecting the steering column, the clock spring will be damaged.

Disconnect the steering column coupling or U-joint from the pinion gear shaft. Scribe alignment marks on the steering shaft and pinion gear shaft so they can be reassembled in the same position. If equipped, disconnect and plug the power steering pressure and return hoses.

Remove the cotter pins and nuts from the tie-rod ends. Using a suitable puller, separate the tie-rods from the steering knuckles.

Support the rack-and-pinion assembly and remove the mounting bolts. In some cases, the rack-and-pinion assembly is mounted to an engine cradle, which may need to be supported and/or lowered in order to remove the rack-and-pinion assembly from the vehicle. Always refer to a vehicle service manual. The manual may also specify that the rack-and-pinion assembly must be removed from one side of the vehicle and/or that the wheels must be removed to facilitate the units removal.

Inspect the rack-and-pinion assembly mount bushings and brackets for wear and damage and replace as required. Loose or damaged bushings or brackets will not allow the unit to be secured to the frame properly and can cause steering problems. Inspect the bellows boots for cracks and tears and replace as necessary.

Before reinstalling the rack-and-pinion assembly, turn the

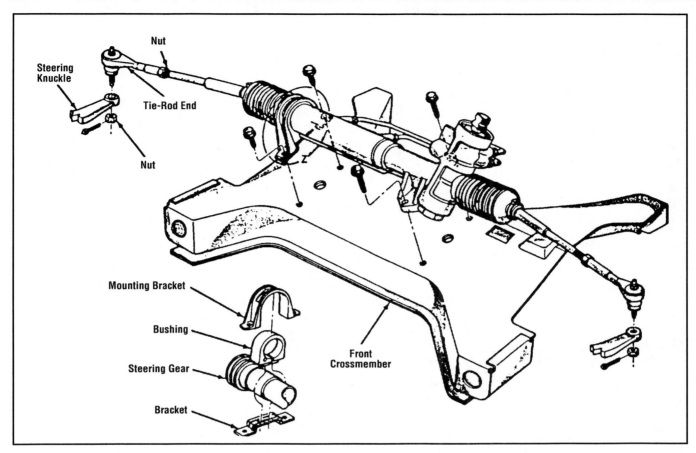

Power rack-and-pinion steering gear mounting. *(Courtesy: DaimlerChrysler Corp.)*

pinion shaft from stop-to-stop, counting the number of turns. Turn the pinion shaft half the number of turns from stop to center the rack.

Align the pinion shaft with the steering shaft and torque all fasteners to specification. If it was necessary to lower an engine cradle to facilitate rack removal, make sure the cradle is properly repositioned. If equipped, bleed the power steering system.

Adjustment

Rack-and-pinion systems sometimes allow for preload and mesh adjustments to compensate for looseness. When they do require adjustment, this is often a symptom of excessive wear in the rack, pinion or housing. Most repair shops will replace a defective or unsatisfactory steering rack rather than attempt to repair it, particularly aluminum-housing steering racks, where scoring of

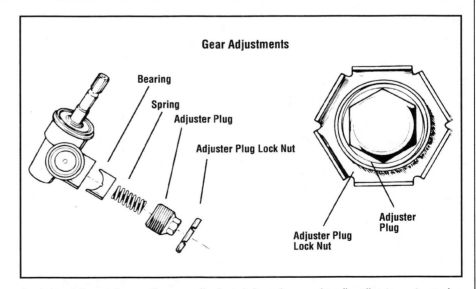

A pinion torque (sometimes called rack bearing preload) adjustment can be made on a rack-and-pinion assembly, but use extreme caution when performing this adjustment so that it is not over-tightened. It is always best to consult a service manual for the proper procedure and specification. Also, it is sometimes difficult to keep the adjuster plug from turning when the locknut is tightened. Always recheck the adjustment after tightening the locknut. *(Courtesy: Federal Mogul/Moog Automotive Division)*

25

TRAINING FOR CERTIFICATION

the aluminum is more common. Rebuilders are equipped to install sleeves to correct this kind of wear, but it is beyond the scope of a general repair shop.

Pinion bearing preload is adjusted either with shims under the pinion end cover or with an adjuster. With the unit removed from the vehicle, turn the pinion shaft with a suitable torque wrench and compare the reading with specifications. Add/remove shims or turn the adjuster to adjust the preload.

Pinion torque is similar to the worm/sector mesh preload in a conventional steering box. It is adjusted either with shims under the rack support cover or by loosening a locknut and turning an adjustment screw or plug. Most of the time, adjustment will call for the pinion shaft to be turned with a suitable torque wrench and the readings compared with specification, and then adjust with shims or by turning the adjuster. However, some manufacturers call for the adjuster to be bottomed in the housing and backed off a specified number of degrees. Always consult the service manual for the vehicle in question for the specific instructions.

Bellows Boot Removal And Installation

Raise and safely support the vehicle and remove the front wheel, as necessary. Loosen the locknut on the inner tie-rod end. Remove the cotter pin, if equipped, and nut from the outer tie-rod end ball stud and separate the tie-rod end from the steering knuckle using a puller.

Remove the outer tie-rod end from the inner tie-rod end. Count the number of turns required to remove the outer tie-rod end from the inner tie-rod end, so that the toe setting will be close after installation. Remove the locknut from the inner tie-rod end. Use suitable cutters to remove the bellows boot retaining clamps, and then remove the boot.

Install the inner boot clamp onto the new bellows boot. Lubricate the end of the steering rack and the inner tie-rod end shaft with suitable grease, then slide the new bellows boot onto the inner tie-rod end until the inner end of the boot is seated in the groove on the steering rack. Make sure that the boot is properly positioned and not twisted or deformed.

Install the outer boot clamp and crimp the clamps using suitable crimping pliers. Install the locknut onto the inner tie-rod end. Thread the outer tie-rod end onto the inner tie-rod end the same number of turns as was required for removal. Attach the tie-rod end to the steering knuckle, torque the stud nut and install a new cotter pin, if equipped. Install the wheel and lower the vehicle. Check the toe setting and adjust as required.

POWER STEERING SYSTEM FLUSHING

Flushing the power steering system is necessary if the fluid is contaminated. Contamination can be caused by a worn pump, steering gear or deteriorated hoses. After repairing the source of the contamination, flush the system as follows:

Raise and safely support the vehicle so the front wheels are off the ground. Disconnect the power steering return hose from the pump and place the end of the hose in a drain pan. Plug the pump return line fitting.

Start the engine and add power steering fluid to the power steering pump reservoir while an assistant slowly turns the steering wheel from stop-to-stop. Do not allow the reservoir to run dry while the engine is running. Continue adding fluid to the reservoir until the fluid coming from the return hose is clear.

Reconnect the power steering return hose to the pump, then fill and bleed the system.

POWER STEERING SYSTEM BLEEDING

Air in the power steering system will cause the fluid to appear foamy and can cause noise and steering problems. Bleeding removes air from the system and is required after any hydraulic system parts have been replaced. Bleed the system as follows:

Run the engine until the power steering fluid reaches operating temperature. Shut the engine off and check the fluid level; add fluid as necessary. Raise and safely support the vehicle so the front wheels are off the ground.

Start the engine. Watch the fluid level in the reservoir as an assistant slowly turns the steering wheel from stop-to-stop. The steering wheel must not be held against the stop for more than

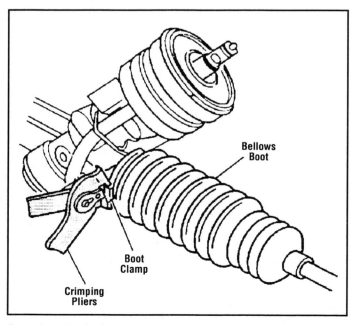

Securing the bellows boot to the steering rack by crimping the inner boot clamp. *(Courtesy: GM Corp.)*

two or three seconds. Add fluid to the reservoir as it runs low; do not allow the reservoir to run dry while the engine is running.

Turn off the engine and check the fluid condition. If the fluid still appears foamy, continue bleeding the system. If the fluid is OK, lower the vehicle and recheck the fluid level.

STEERING LINKAGE

The two most common steering linkage designs in use today are the parallelogram steering linkage used with a conventional steering box, and the rack-and-pinion system. In the parallelogram design, a center link is connected to the pitman arm, which is attached to the steering box, and a frame mounted idler arm, forming a parallelogram. Steering motion from the center link is transferred to the steering knuckles by the tie-rod assemblies. In the rack-and-pinion system, the steering rack takes the place of the center link. The tie-rod assemblies are attached to the ends of the rack and connected to the steering knuckles.

There are two common variations of pitman arm/center link/parallelogram steering geometry used on some four-wheel drive trucks. These variations are used mainly because their geometry leaves more space for the front driveline components.

One system is known as the Cross Steer system. In such a system, the pitman arm pulls and pushes a steering link called the drag link, which is variable in length just like a tie-rod. The drag link works the left front wheel steering arm directly, and there is a second, forward-angled 'steering arm' on the left steering knuckle connected by a long adjustable steering link to the steering arm of the right wheel. Single adjusters are on the crossover link and the drag link, making it sometimes necessary (as on some small front-wheel drive vehicles) to remove and reposition the steering wheel to get it straight at the end of an alignment.

The other system is known as the Haltzenberger system. This system utilizes a connecting link that is attached through ball studs to the pitman arm and a tie-rod. In this system, the pitman arm controls the side-to-side movement of the tie-rod. The tie-rod is equipped with adjustable tie-rod ends that connect to the steering knuckles. The connecting link is not adjustable.

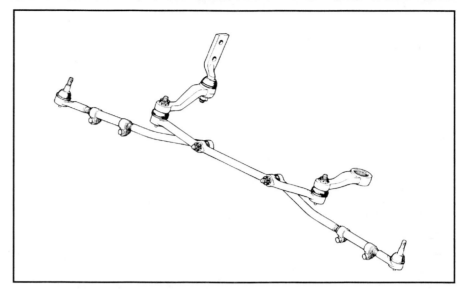

A parallelogram steering system.

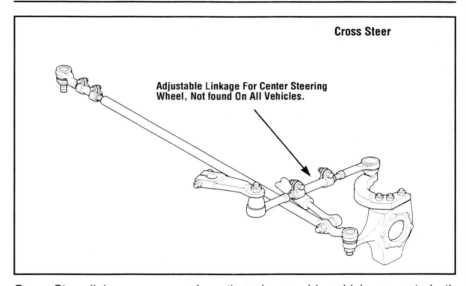

Cross Steer linkage uses one long tie-rod assembly, which connects both the right and left steering arms. A drag link then connects the left steering arm to the steering gear. On some vehicles, an adjustable tie-rod assembly replaces the drag link. This allows the steering wheel to be centered.
(Courtesy: Federal Mogul/Moog Automotive Division)

Steering Linkage Service

Most steering linkage is connected with small ball joints. The ball fits into a socket housing that is connected to one link and the tapered stud on the other end of the ball fits into a tapered hole in the connecting link. The stud is secured with a castle nut and cotter pin or a locknut. A dust

27

TRAINING FOR CERTIFICATION

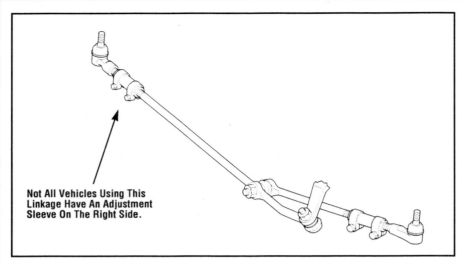

Not All Vehicles Using This Linkage Have An Adjustment Sleeve On The Right Side.

A Haltzenberger linkage uses a pitman arm and two linkages to connect the pitman arm to the steering arms. The left side linkage has an adjustment sleeve to perform toe adjustments. Some vehicles are equipped with a three-piece linkage on the right side. This allows the steering wheel to be centered, as toe is set. If the steering linkage does not have an adjustment sleeve on the right side, the steering wheel is centered by removing and installing it after total toe is set.
(Courtesy: Federal Mogul/Moog Automotive Division)

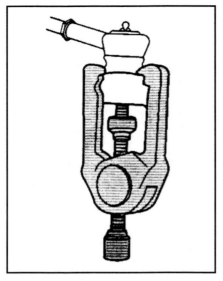

Using a puller to separate a steering linkage ball joint stud from the tapered hole in a steering system component. *(Courtesy: GM Corp.)*

cover is installed over the ball and socket assembly to keep dirt out and lubricant in. Most steering linkage ball joints are permanently lubed, however, some are equipped with grease fittings to allow periodic lubrication.

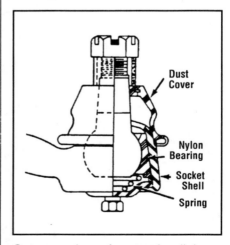

Cutaway view of a steering linkage ball joint.

Steering linkage ball joints should be replaced if there is any looseness in the ball/socket or stud/tapered hole junctions, if the stud turning effort is excessive (refer to Steering System Diagnosis in the Steering Units section), or if the dust boot is cracked or torn. In order to disconnect a steering linkage ball joint, remove the cotter pin and nut or locknut and separate the tapered stud from the tapered hole. Using a puller designed for this purpose is the preferred method of separation, especially if the ball joint is not to be replaced. Separating the stud from the tapered hole by driving a ball joint separator, also known as a 'pickle fork', between the joint will destroy the dust boot, making joint replacement mandatory.

Before installing a new steering link, make sure there is no damage to the tapered hole so that the stud will seat properly. Torque the locknut or castle nut to specification and install a new cotter pin. If the cotter pin hole does not align with the nut castellation, tighten the nut further just enough to install the cotter pin. Never loosen the nut to install the cotter pin.

PITMAN ARM

The pitman arm, which is connected to the output or sector shaft of the steering gearbox, delivers the driver's input to the working parts of the steering linkage. Most pitman arms do not have any moving parts and are known as nonwear pitman arms. This type of pitman arm should only require replacement if it is physically damaged in an accident or if the tapered hole for the corresponding steering linkage ball joint stud is damaged. The other type of pitman arm has a ball joint to connect to the steering linkage, and is known as a wear pitman arm. This type of pitman arm should be replaced if there is any wear in the joint.

The pitman arm is mounted on the tapered spline of the sector shaft. After removing the nut and lockwasher from the sector shaft, remove the pitman arm using a suitable puller. Most pitman arms are indexed to the sector shaft, but if not, scribe alignment marks prior to removal so the pitman arm can be reinstalled in the same position. Use a puller or ball joint separator to disconnect the pitman arm from the steering linkage.

When installing the pitman arm, index the pitman arm with the sector shaft or align the marks made at removal. Torque the pitman arm-to-sector shaft nut to specification. Torque the steering linkage ballstud locknut or castle nut to specification and install a new cotter pin.

CENTER LINK

A bent or damaged center link can cause front-end shimmy, steering pull and incorrect toe, which in turn can cause tire wear. Inspect the center link for worn linkage joints or other damage.

The center link is replaced by disconnecting and connecting the necessary steering linkage ball joints, as explained earlier in this section.

When installed, the center link should be centered when the front wheels are in the straight-ahead position, the pitman arm and idler arm should point forward, and the center link should be parallel with the ground. If the center link is not in this position, a bump steer condition will result. Check the pitman arm and idler arm for damage or incorrect positioning, which could cause this condition.

IDLER ARM

The idler arm is the same length and set at the same angle as the pitman arm. Its function is simply to hold the right end of the center link level with the left end, which is moved by the pitman arm. The weak point on the idler arm is the bushing: if that wears, it can allow the arm to pivot up and down as well as side-to-side. That up-and-down play can allow changes in the relative toe between the two front wheels.

Inspect the idler arm by grasping the center link as close to the idler arm as possible and move it up-and-down. Look at the right front tire and note any toe changes. Excessive vertical movement in the idler arm socket can cause undesirable changes in toe.

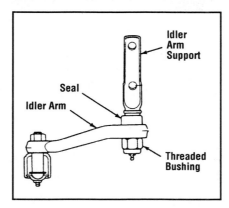

A worn idler arm can cause front end shimmy and shake. The idler arm should securely hold the center link parallel to the ground.

An idler arm is removed by disconnecting the steering linkage ball joint and removing the bolts that attach it to the frame. Some idler arms have replaceable bushings, but most idler arms are replaced as a unit.

Some idler arms are adjustable through slots in the frame where the idler arm connects, so it is important to mark its mounting location prior to its removal so it can be reinstalled in the same location. If these adjustable idler arms are installed incorrectly, the center link will not be parallel and a bump steer condition will result.

TIE-RODS

The tie-rods connect the center link or steering rack with the steering knuckles. Most tie-rod assemblies on conventional steering vehicles consist of two tie-rod ends connected with an adjustment sleeve. On rack-and-pinion steering equipped vehicles, the inner tie-rod end pivots in a socket attached to the end of the rack. The outer tie-rod end threads onto the inner tie-rod end and is held in place with a locknut.

Worn tie-rods can cause front-end shimmy and incorrect toe, which in turn can cause tire wear. Inspect the tie-rods by grabbing the tie-rod near its ball joint and use moderate force to move it up and down and in and out. Also check for movement where the tie-rod end threads into the adjustment sleeve. If the tie-rod has excessive movement or causes significant changes in toe, it should be replaced. The tie-rod should also be replaced if the ball joint is binding.

All Except Rack-And-Pinion Steering Gear Inner Tie-Rod End

To replace a worn tie-rod end, loosen the adjustment sleeve clamp bolt or locknut and disconnect the ball stud from the steering knuckle. As you unscrew the tie-rod end from the adjustment sleeve or inner tie-rod end, count the number of turns required for removal. Thread the new tie-rod end into the sleeve or tie-rod end the same number of turns, so the toe setting will be close. Check and adjust the toe after the tie-rod end is installed.

Tie-rod adjustment sleeves and tie-rod ends have left- and right-hand threads to allow adjustment of the tie-rod's length. These adjustment sleeves are the main way the toe is set. Some tie-rod ends themselves are hollow, and the adjuster is the solid rod, but they work in exactly the same way.

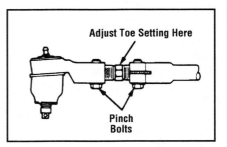

Whether the toe is adjusted with a turn-bolt as shown here, or an adjustment sleeve, make sure you position and tighten the clamps when you finish adjustments.

Adjustment sleeves have locking clamps at either end. In general, you can set these clamps at any position that makes tightening the clamp convenient, but the tie-rod ball stud should be centered in the ball socket prior to

TRAINING FOR CERTIFICATION

tightening these clamps upon installation of the tie-rod or at the conclusion of a toe adjustment. Additionally, some manufacturers caution you not to align the clamp opening with the slot in the adjusting sleeve. Make sure that the tie-rod adjusting sleeve bolts do not contact the vehicle's suspension during its normal course of travel. Check this by turning the steering wheel from stop-to-stop while the vehicle is supported on the lift. If the tie-rod sleeve adjusting bolts do not contact the vehicle's suspension under this condition, it will most likely be OK while the vehicle is being driven as well.

A few small front-wheel drive vehicles have an adjustable tie-rod on only one side, as do a few four-wheel drive vehicles. This does not prevent setting the toe properly, but it may present a difficulty in setting the steering wheel straight. On most of these vehicles, the steering wheel can be set in more than one position on the column shaft, and one position will be close to correct. There are adjustable sleeves available in the aftermarket for most of these vehicles, which will allow for a perfect orientation of the steering wheel.

Many tie-rod ends have grease fittings; others are permanently lubed. A few, particularly on some Fords, use a Rubber-Bushed Solid (RBS) tie-rod end. A RBS tie-rod is a design that uses a solid bushing as opposed to a ball and socket design. When replacing RBS tie-rods, it is important to set the steering to straight-ahead before tightening the nut on the tie-rod end ball stud. If this step is not taken, the steering will always have a memory steer condition, which is a tendency to turn back to the position it was in when the tie-rod end ball stud nut was tightened. To avoid this condition and to help ensure the vehicle steering always tends to point straight-ahead, it is recommended that RBS tie-rod ends be replaced in pairs.

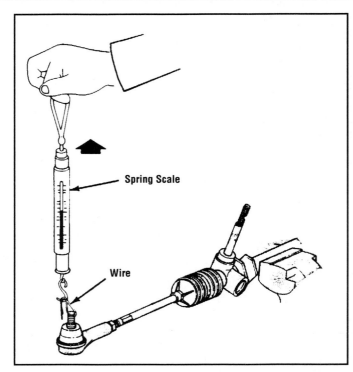

Measuring tie-rod articulation effort with a spring scale. (Courtesy: Ford Motor Co.)

Rack-And-Pinion Steering Gear Inner Tie-Rod End

Replacement of a steering rack inner tie-rod end is necessary if it is loose or fails an articulation (pivot) effort test. To measure the effort required to pivot the inner tie-rod end, the outer tie-rod end must be removed from the steering knuckle. Attach a spring scale and pull down on the tie-rod. Compare the reading with specification.

Consult a service manual to see if removal of the rack is necessary for inner tie-rod end replacement. With the outer tie-rod end disconnected from the steering knuckle, count the number of turns required to remove the outer tie-rod end from the inner tie-rod end. This is done so that the toe setting will be close after installation. Use suitable cutters to remove the bellows boot retaining clamps, and then remove the boot.

Various methods are used to secure the inner tie-rod end to the steering rack. Some inner tie-rod end sockets are integral with the tie-rod end; the whole unit threads into the rack and is secured with lock pins or lock washers. These units are preset and lubed for life. On other designs, the socket and the inner tie-rod end ball must be lubed and the tie-rod end attached with a locknut. The locknut is then either torqued or tightened until a specified force needed to pivot the tie-rod is reached, measured with a spring scale. The locknut is then pinned in place.

Install the bellows boot with new clamps. Thread the outer tie-rod end onto the inner tie-rod end the same number of turns as was required for removal. Attach

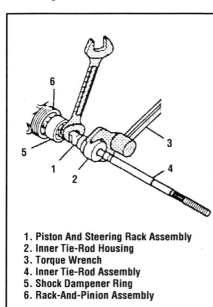

1. Piston And Steering Rack Assembly
2. Inner Tie-Rod Housing
3. Torque Wrench
4. Inner Tie-Rod Assembly
5. Shock Dampener Ring
6. Rack-And-Pinion Assembly

Inner tie-rod end installation. (Courtesy: GM Corp.)

the tie-rod end to the steering knuckle, torque the stud nut and install a new cotter pin, if equipped. Check the toe setting and adjust as required.

STEERING LINKAGE DAMPER

Some vehicles are equipped with a steering damper attached to the steering linkage. The steering damper's purpose is to control road shock and vibration. A steering damper looks like and functions much like a shock absorber.

Service to steering dampers is usually limited to replacement. Inspect the unit for leaking fluid and worn mounting bushings and joints, and replace as required.

NOTES

NOTES

TRAINING FOR CERTIFICATION

SUSPENSION SYSTEMS DIAGNOSIS AND REPAIR

The suspension system serves two purposes, to keep the vehicle in control by keeping the wheels in contact with the road and to provide a smooth ride for the vehicle's passengers. There are many suspension system designs, but all employ a system of springs to support the vehicle and absorb bumps, and shock absorbers to control the action of the springs.

FRONT SUSPENSIONS

There are several front suspension systems used on cars and light trucks. The following are descriptions of the most common designs.

MACPHERSON STRUT SUSPENSION

This system is built around the strut, which is a combination shock absorber and spring. In this system, the strut assembly acts as a suspension member. The top of the strut acts as a top mounting point for the suspension system. This top mount supports the vehicle load and serves as a pivot point for the strut when the wheels are turned.

The lower end of the strut is attached to the steering knuckle, which is attached to a single lower control arm. The control arm can attach to the frame at one or two inner pivot points. If only one inner pivot point is used, it is generally located fore-and-aft with a strut rod or stabilizer bar. The control arm is designed to handle side loading. In some cases, the steering arm may be connected to the strut assembly. This occurs when the vehicle is equipped with a center take off-type steering rack. In these cases, the strut also acts as the steering knuckle.

A stabilizer bar connects the left and right lower control arms, usually through end links. One variation has a cast steering knuckle with a drive axle passing through it. When used in a front drive system, the cast knuckle replaces the integral spindle of a non-drive knuckle. The principal advantage of this system is its simple design and lighter weight, allowing vehicle manufacturers to make greater use of the space available in the engine compartment and for powertrain location.

MODIFIED STRUT SUSPENSION

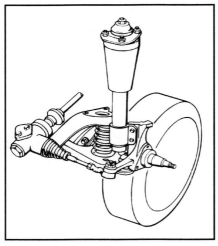

A front suspension system with modified struts.

This system has one control arm and a modified strut design. The coil spring sits on the control arm and is supported at the top by a fixed point, usually a frame member. The control arm can resemble the wishbone type used in SLA (Short/Long Arm Suspension) systems or is sometimes a modified beam-type. It is attached to the frame by one or two pivot points, depending on the type of arm used. In this system, the stabilizer bar, connecting the control arms, handles body sway but not fore-aft location.

MODIFIED DOUBLE WISHBONE

The modified double wishbone is another type of strut suspension that is becoming more common. It combines the space saving benefits of a strut suspension system with the ability of the parallel arm suspension to ride low to the ground. This allows for a more aerodynamic hood-line.

With this design, the lower portion of the strut forms a wishbone shape where it attaches to the lower control arm. Unlike

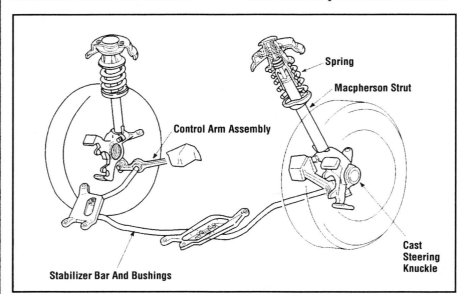

A MacPherson strut front suspension with cast steering components.

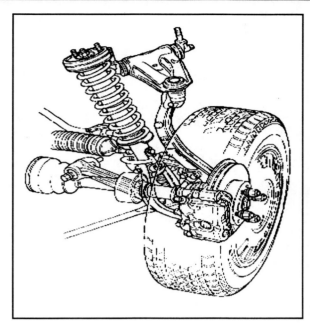

A modified double wishbone front suspension.
(Courtesy: Tenneco Automotive)

other struts, the double wishbone does not rotate when the wheels turn. Instead, the entire spindle assembly rotates on the upper and lower ball joints much like a parallel arm suspension. Since the strut does not rotate, the upper mount does not need a bearing. Instead, a hard rubber bushing replaces the bearing and helps isolate road shock.

SHORT/LONG ARM SUSPENSION (SLA)

This system, also known as the double wishbone, has a coil spring located between the lower control arm and a frame mounting point, such as a crossmember. There was one exception to this with some '60s Ford vehicles that had the spring mounted on the upper arm. The design is so named because the lower arm is longer than the upper arm.

The control arms are attached to the vehicle frame with rubber torsilastic bushings. Rubber bushings are preferred because they do not require lubrication, and will reduce minor road noise and vibrations. Torsilastic refers to the elastic nature of rubber to allow movement of the bushing in a twisting plane. Movement is allowed by the twisting of the rubber.

The outer sleeve of the bushing is press fit into the control arm, while the inner sleeve is locked to the control arm pivot shaft. The rubber must twist to allow movement of the control arm. This twisting action of the rubber will provide resistance to movement.

The outer end of the control arm is connected to the steering knuckle with a ball joint. Ball joints are simple connectors, which consist of a ball and socket. The ball and socket assembly forms the steering axis for the suspension system.

A stabilizer bar, often called a sway bar, connects the two lower arms. When the suspension at

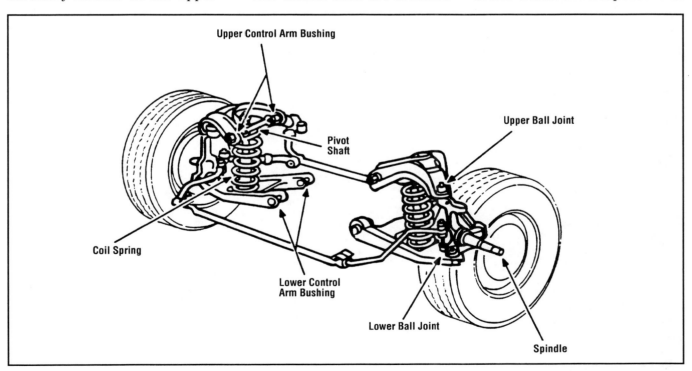

Short/Long Arm (SLA) type suspensions are a common kind of independent suspension system. The triangular sheet metal control arms are much lighter than a solid axle, and the geometry of different length arms keeps wheel camber and tread-patch lateral movement in control.
(Courtesy: Hunter Engineering Co.)

TRAINING FOR CERTIFICATION

one wheel moves up and down, the bar transfers the movement to the other wheel. For example, if the left wheel drops into a pothole, the sway bar transfers the movement to the right wheel. Thus the sway bar creates a more level ride and reduces vehicle sway or body roll during cornering.

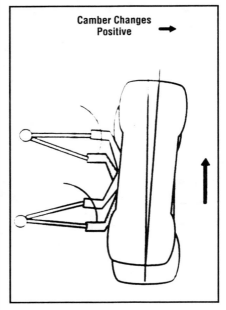

A jounce movement on a SLA type suspension causes the wheel to move upward and results in a positive camber change.
(Courtesy: Hunter Engineering Co.)

The main advantage of the SLA system is its ability to minimize toe-in during turns. Using a shorter upper arm causes a slight camber change as the vehicle travels through jounce and rebound. While this may sound bad, it actually is not. If both arms were the same lengths, a track change would occur causing the tire to travel sideways. The tires would then scrub the pavement, causing tire wear and handling problems.

The double wishbone was the setup of choice for vehicle manufacturers through the 1970s. However, with the advent of the strut and single lower arm suspension, the SLA suspension fell

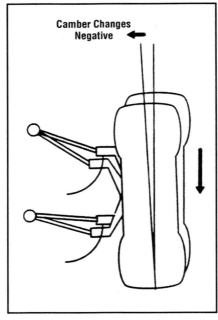

A rebound movement on a SLA type suspension causes the wheel to move downward and results in a negative camber change.
(Courtesy: Hunter Engineering Co.)

out of favor. The simpler design and lighter weight of the new system allowed manufacturers to make greater use of room in front end design. Ironically, the double wishbone is now returning as the design of the future.

Most auto manufacturers are redesigning the venerable double wishbone with new technologies in metal castings and lightweight components. The new SLA front ends allow a lower profile while providing the increased strength and resistance to roll provided by the upper/lower arm geometry.

TWIN I-BEAM SUSPENSION

The twin I-beam is a type of independent front suspension that was used on light trucks, vans and four-wheel drive vehicles due to its load carrying ability. Although it is similar to the solid axle in many ways, it was designed to improve ride and handling. The improvement is achieved through the independent movement of each I-beam, which allows improved handling characteristics without the windsteer common to other independent suspensions on trucks.

The twin I-beam suspension consists of two short I-beams supported by coil springs, and the steering knuckles/spindles, which are connected by kingpins or ball joints. The inner end of the axle connects to the vehicle frame through a rubber bushing. A radius arm also connects to the

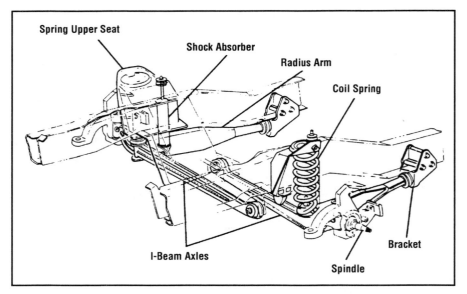

The Ford twin I-beam suspension has some of the unsprung weight advantages of independent suspension, but retains the strength and simplicity of a solid axle design.

frame through rubber bushings. This arm controls wheelbase and caster.

While the twin I-beam design was an improvement over the solid axle, it still has some flaws. For example, with the twin I-beam, the camber and track change as the wheels move up and down, creating tire wear.

SOLID AXLE SUSPENSION

One of the longest established forms of suspension is the solid axle, connected to the vehicle by leaf or coil springs. It is simple and strong and requires little maintenance. Realignment is necessary only if parts bend or break. Today, solid axles are used only on some 4WD trucks.

When equipped with leaf springs, the axle is located by the springs. No other components are necessary, but for handling and control, a Panhard rod or stabilizer bar is used.

With coil springs, control arms or radius rods are used to locate the axle.

The downside of solid axles is that they provide a rougher ride than independent suspension. Solid axle vehicles are also subject to wind-steer and steering wander.

TORSION BARS

There are no leaf or coil springs in a torsion bar suspension. Instead, a torsion bar supports the vehicle weight and absorbs the road shock. Actually the torsion bar performs the same function as a coil spring: it supports the vehicle's weight. The difference is that a coil spring compresses to allow the tire and wheel to follow the road and absorb shock, while a torsion bar uses a twisting action. Other than this difference, however, the two types of suspension construction are much the same.

The torsion bar connects to the upper or lower control arm at one end, and at the other end connects to the frame. It can be mounted longitudinally, front to rear, or transversely, side-to-side. Unlike coil springs and leaf springs, torsion bars can be used to adjust suspension ride height. However, torsion bars are not normally interchangeable from side-to-side. This is because the direction of the twisting or tor-

Torsion bar suspensions support the vehicle's weight on longitudinal springs that twist as the suspension works.
(Courtesy: Tenneco Automotive)

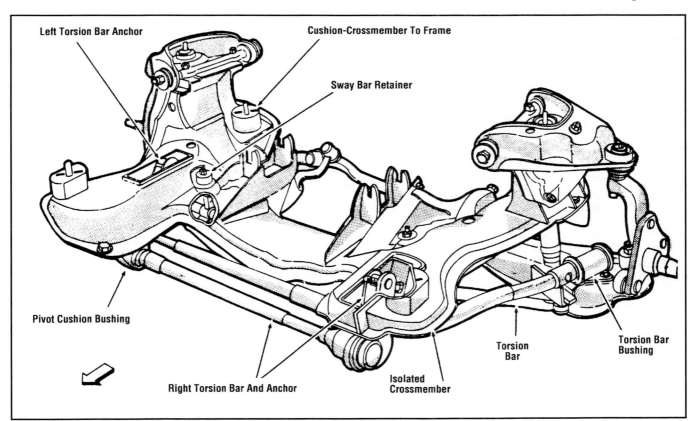

Not all torsion bar suspensions use longitudinal torsion bars. This suspension uses two transverse mounted torsion bars.

37

TRAINING FOR CERTIFICATION

sion is not the same on the left and right sides.

Front Suspension System Diagnosis

The most common suspension system complaints are noise, abnormal tire wear, poor ride quality, steering pull, front end shimmy and excessive body roll. These problems are most often caused by worn bushings, ball joints and shock absorbers.

Bushings are usually made of rubber and when the rubber deteriorates, banging or clunking noise can result from metal-to-metal contact. Worn control arm bushings can cause shimmy, tire wear and steering and wheel alignment problems. Worn radius arm/strut rod bushings can cause the vehicle to pull to one side when braking or jump to one side when hitting a bump. Worn sway bar bushings can cause increased body roll in turns. Bushings can also become dry and hard from age, resulting in a squeaking noise.

Ball joints wear from prolonged use or lack of lubrication. Worn ball joints allow excessive movement between the steering knuckle and control arm. This can result in a clunking noise and cause shimmy, and steering and wheel alignment problems.

When shock absorbers are worn, they cannot adequately dampen coil spring oscillations, resulting in poor ride quality and handling problems. A shock absorber can also be the source of a banging noise if it is loose in its mounts or its mounting bushings are worn.

When inspecting the front suspension, look for the obvious first: leaking shock absorbers, worn bushings, broken springs, bent or broken sway bar links and bent control arms. Also, keep in mind that since suspension and steering are interrelated systems, the steering system as well as wheel bearings and tires should also be inspected. Check the tires for proper inflation and abnormal wear patterns that could be caused by incorrect wheel alignment, which in turn could be caused by suspension wear. Make sure the wheel bearings are properly adjusted.

Give a firm push down at each corner of the bumper and see how many times the vehicle rebounds, listening for noise as the vehicle moves up-and-down. If the vehicle rebounds more than twice, the shock absorbers are worn. Noise may be evidence of a loose shock absorber or worn control arm bushings. A wheezing or creaking sound may be caused by a shock or strut with dried out seals. Another sign of worn shocks or struts is an irregular, choppy or scalloped wear pattern on the tires. This wear pattern indicates that the shocks or struts are weak and are not keeping the tires in contact with the road surface.

Check for obvious vehicle ride height problems indicated by a list to one side, or front or rear end sag. If in doubt, measure the ride height and compare with manufacturer's specifications. Incorrect ride height is usually caused by worn or broken springs. Ride height can be adjusted on vehicles with torsion bars.

To inspect ball joints and control arm bushings for wear, the tension must be removed at the joint or bushing. If the coil spring is positioned between the frame and lower control arm, support the vehicle so the weight is on the lower control arm. If the coil spring is between the frame and upper control arm, or if the spring is on the strut, support the vehicle so the weight is on the frame and allow the suspension to hang free.

With the tension removed, check for excessive control arm bushing wear by moving the control arms vertically and horizontally, watching for excessive movement of the arm in relation to the shaft. Any movement warrants bushing replacement. Check for ball joint wear by moving the wheel up-and-down to check for axial movement (it may be necessary to use a prybar for leverage) and in-and-out for radial movement. Compare the amount of movement with manufacturer's specifications to determine if ball joint replacement is necessary.

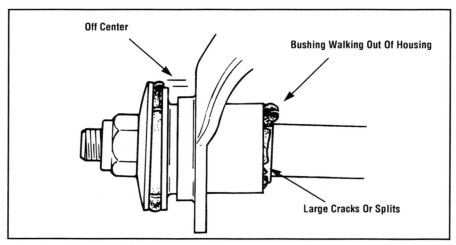

Visually inspect bushings for distortion, large cracks or splits, off center condition and indication of excessive movement (e.g. polished or shiny areas). Visual observation of some bushings may be difficult due to powertrain and sheet metal configurations. In these cases, inspect the bushings by jouncing the vehicle. If there is any noise present, there is a good chance that the bushing on the moving component(s) is separated internally or rotating on its pivot shaft or bolt.
(Courtesy: Federal Mogul/Moog Automotive Division)

After visual inspection, if possible, road test the vehicle. Even if components like shocks and struts appear OK, they still may not perform properly. If the vehicle dives badly under braking, seems to 'float' at highway speeds, or leans badly in turns and then does not settle down immediately after the turn, new shocks or struts are required.

Front Suspension System Service

CONTROL ARMS AND BUSHINGS

Control arms connect the vehicle frame to the steering knuckles and allow the up-and-down movement of the wheels. The control arm is connected to the steering knuckle with a ball joint and pivots at the frame on bushings.

If the coil spring is located between the vehicle frame and lower control arm and the upper control arm is to be removed, support the vehicle on jack stands positioned under the lower control arm close to the ball joint. Disconnect the upper ball joint from the steering knuckle, remove the control arm mounting bolts at the frame and remove the control arm. Keep track of the position of any alignment shims.

To remove the lower control arm, use a suitable spring compressor to compress the coil spring. Disconnect the sway bar link and strut rod from the control arm, as necessary. Disconnect the lower ball joint from the steering knuckle, remove the control arm mounting bolts at the frame and remove the lower control arm.

Control arm bushings are usually replaced using a press and suitable press fixtures. When the control arm is removed from the vehicle, inspect the control arm bump stops for wear and replace as necessary. When installing the control arm, torque all fasteners to specification and install a new cotter pin in the ball joint stud. The control arm bushing fasteners should not be tightened until the weight of the vehicle is on the suspension and the vehicle is at normal ride height.

RADIUS ARMS/STRUT RODS

Strut rods are used to brace the lower control arm and are used in suspension designs where the lower control arm attaches to the frame at only one point. Radius arms are used to brace an axle or I-beam.

Strut rod replacement usually only involves supporting the vehicle so the weight is off the suspension, unbolting and removing the strut rod and bushings, replacing the rod and/or bushings as required and torquing all fasteners. Radius arm replacement is more involved. After the vehicle has been raised and supported at the frame, the axle or I-beam must be supported and the wheel removed. The sway bar link is then disconnected and the shock absorber removed.

Lower the axle or I-beam and remove the coil spring. It may be necessary to remove the brake caliper to prevent damage to the brake hose. Unbolt the radius arm and remove it from the vehicle.

Inspect the radius arm and bushings for wear and damage and replace as required. Torque all fasteners to specification.

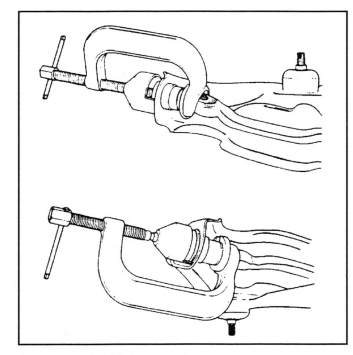

Control arm bushing removal.
(Courtesy: Ford Motor Co.)

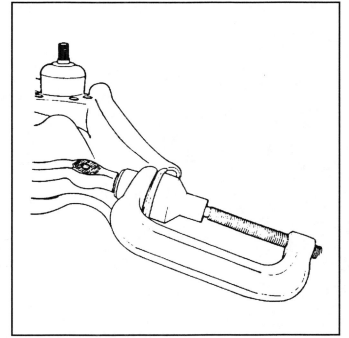

Control arm bushing installation.
(Courtesy: Ford Motor Co.)

TRAINING FOR CERTIFICATION

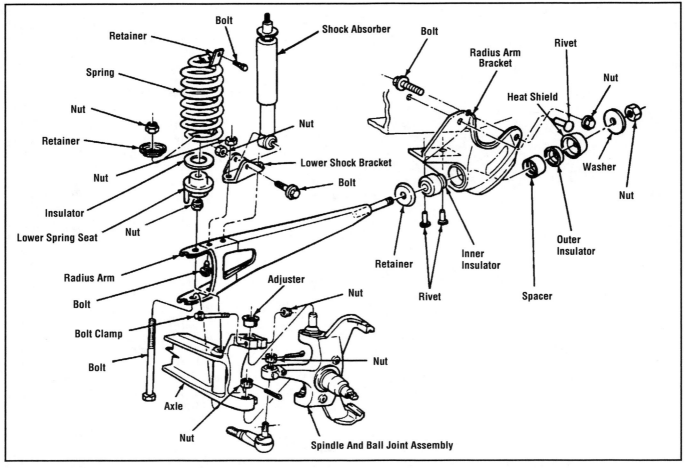

Radius arm installation. *(Courtesy: Ford Motor Co.)*

BALL JOINTS

The ball joint provides a pivot point, allowing the steering knuckle to move up-and-down as well as turn in response to steering input. The ball fits into a socket housing that is attached to the control arm and the tapered stud on the other end of the ball fits into a tapered hole in the steering knuckle. The stud is secured with a castle nut and cotter pin. Ball joints on strut suspensions often use a straight rather than tapered stud. The stud is held in a pinch clamp at the bottom of the strut, and often the stud has a slot that lines up with the pinch clamp throughbolt. A dust cover is installed over the ball and socket assembly to keep dirt out and lubricant in. Many ball joints are permanently lubed, however, some are equipped with grease fittings to allow periodic lubrication.

Check for ball joint wear as described in Front Suspension System Diagnosis. Some ball joints include a wear indicator as part of their original design. If there is a shoulder into which the grease fitting threads, the ball joint is considered worn out once

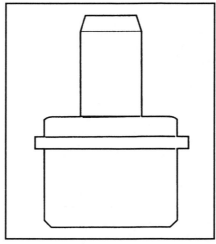

A straight stud/pinch bolt ball joint.

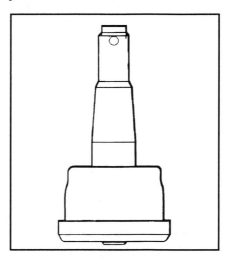

A tapered stud ball joint.

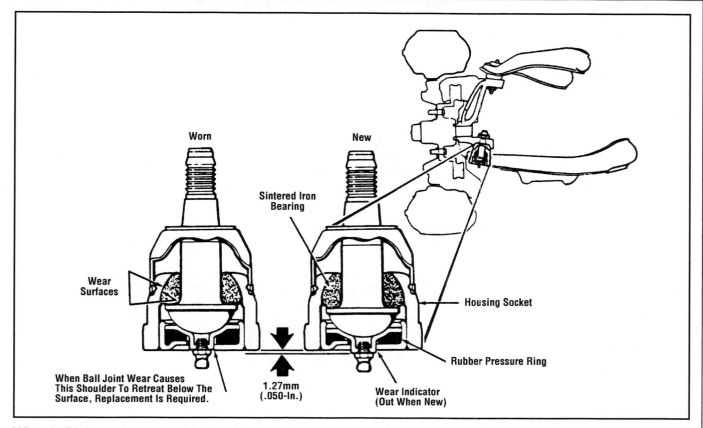

When ball joint wear causes this shoulder to retreat below the surface, replacement is required.

the shoulder recedes so far into the hole that its outer surface is level with the metal surrounding it. For such joints, replacement is in order regardless of the level of measurable play. A ball joint also must be replaced if its dust cover is broken. Once grit can enter the joint, wear accelerates rapidly, and the joint can fail.

Most ball joints fail due to wear in the ball and socket area, however, if the ballstud has broken, it is possible that the tapered hole has become distorted. You can check this by trying the new ball joint stud in the hole: if there is any free-play or if the new tapered stud can rock in the hole, the hole is rounded out. If this has occurred, the steering knuckle must be replaced.

To replace a ball joint, raise and support the vehicle, then remove the wheel. Support the vehicle under the control arm or frame so the ball joint is not under tension. It may be neces-

sary to use a spring compressor to compress the coil spring.

Remove the cotter pin and loosen the castle nut. Separate the tapered stud from the tapered hole using a puller designed for this purpose or by driving a ball joint separator, also known as a 'pickle fork', between the joint and steering knuckle. Keep in mind that using a ball joint separator will destroy

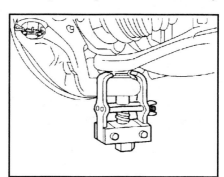

Separating the ball joint from the steering knuckle using a puller. *(Courtesy: Ford Motor Co.)*

the dust cover, making joint replacement mandatory.

Ball joints can be threaded, pressed or riveted onto the control arm. If the ball joint is threaded, remove it using a socket wrench. If the ball joint is pressed into the control arm, a ball joint press should be used to remove it from the control arm. Riveted ball joints must have the rivet heads drilled or chiseled away to remove the ball joint.

A new threaded ball joint should be torqued to specification. Press a new press-fit ball joint into position using a ball joint press. Attach a replacement for a riveted ball joint with the hardware supplied with the kit, torquing the bolts/nuts to the specifications listed on the instructions in the kit.

Insert the ball stud into the steering knuckle and install the castle nut. Torque the castle nut to specification and install a new cotter pin. If the cotter pin hole

41

TRAINING FOR CERTIFICATION

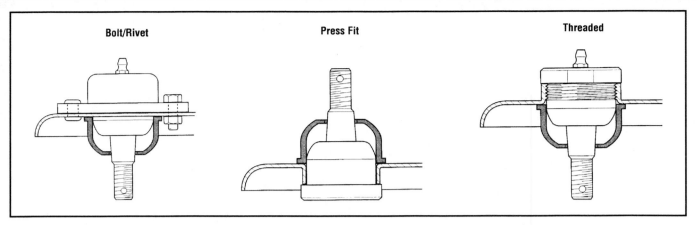

Different types of ball joint mounting.

does not align with the nut castellation, tighten the nut further just enough to install the cotter pin. Never loosen the nut to install the cotter pin.

Install the grease fitting, if provided, and lubricate the ball joint. Install the wheel and lower the vehicle.

NON-INDEPENDENT FRONT AXLE

A non-independent, or solid, front axle can be either the differential and axle housing assembly on a 4WD truck or the beam axle found on older vehicles. Damage to the axle that causes misalignment can cause abnormal tire wear and vehicle tracking and handling problems.

Inspect the axle assembly for bending, warpage and misalignment. The damage may not be evident and a problem may only reveal itself through the aforementioned tire wear, and tracking and handling problems. In this case front wheel alignment must be checked against specifications to find the cause and extent of the damage. Adjustment to bring alignment within specifications may not be possible or the amount of adjustment possible may not be sufficient, making axle replacement necessary.

KINGPINS

Kingpins are used on some trucks with solid axle or twin I-beam suspensions. The kingpin attaches the wheel spindle to the axle or I-beam and provides the pivot for the wheels to be turned. The kingpin is held solidly in place with a pin on the axle or I-beam and the spindle turns on bushings. Inspect for wear to the bushings and kingpin by raising the vehicle so the wheels are off the ground, and shaking the wheel while observing the spindle in relation to the axle or I-beam. Compare any movement with allowable specifications to determine whether bushing replacement is warranted.

To replace the kingpin and/or bushings, raise and support the vehicle so the wheels are off the ground. Remove the wheel and the brake components. Next, disconnect the tie-rod end from the spindle.

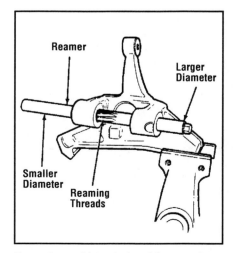

Reaming a kingpin bushing to size.
(Courtesy: Ford Motor Co.)

Remove the lockpin nut and remove the lockpin from the axle/I-beam. Remove the upper and lower kingpin plugs from the spindle, and then use a suitable punch to drive the kingpin from the spindle and axle/I-beam. Examine the kingpin for wear to determine if replacement is necessary. Remove the spindle and thrust bearing.

The kingpin bushings are driven out of the spindle using suitable drivers. After the new bushings are installed, they are reamed or honed to provide the proper clearance.

Position the spindle on the axle/I-beam with a new thrust bearing and seals. Some vehicles use shims between the spindle and axle/I-beam so a measurement must be taken at this time and the proper shims installed. Lubricate the kingpin and bushings with grease and install the kingpin, aligning the lockpin notch with the hole in the axle/I-beam. Install the lockpin and torque the nut to specification.

Install the kingpin plugs on the spindle and lubricate the kingpin and bushings with grease at the grease fittings. Connect the tie-rod end and install the brake components and wheel. Lower the vehicle.

STEERING KNUCKLE

The steering knuckle connects the upper and lower control arms or the strut and lower control arm. On rear wheel drive vehicles, it usually incorporates the front wheel spindle and on front wheel drive vehicles it has an opening where the halfshaft passes through. A steering arm is attached to the steering knuckle, where the tie-rod end is connected.

A steering knuckle should be replaced if it is bent, cracked or if there is damage to the spindle or the tapered ball stud holes. Steering knuckle removal is sometimes required for bearing service on front-wheel drive vehicles.

To remove the steering knuckle, raise the vehicle and support it so the wheels are off the ground. Remove the wheel and the brake components. Disconnect the tie-rod end from the steering arm.

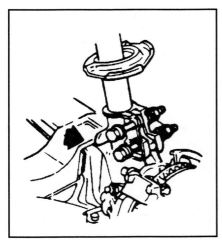

The strut-to-steering knuckle mounting bolts on this vehicle are also used to adjust camber. The strut and knuckle must be scribed prior to steering knuckle removal so the original camber setting is maintained.
(Courtesy: GM Corp.)

Support the suspension so that tension is removed from the ball joint(s). Support the steering knuckle and separate it from the lower ball joint. Separate the knuckle from the upper ball joint or strut. On some strut vehicles, the strut-to-knuckle bolts also are used for camber adjustment. In this case, mark the position of the fasteners or the strut in the knuckle so that the camber setting will be the same after installation.

Install the steering knuckle and torque all fasteners to specification. Connect the tie-rod end, install the brake components and wheel, and lower the vehicle. Check the front-wheel alignment.

COIL SPRINGS

Coil springs are positioned between the axle and a fixed point above the axle, such as the frame, a mounting pad, a control arm or on a strut. The spring is designed to compress to a certain point when loaded with the vehicle's unladed weight. A compression rate is engineered into the spring to allow for controlled compression when the vehicle is loaded, and to allow controlled rebound when the suspension travels over an uneven surface. As springs wear, the compression rate changes and handling characteristics will be affected. When replacing springs, they should always be replaced in axle sets.

There is a considerable amount of energy stored in a compressed coil spring. If you must remove a coil spring, either because it's worn or because it is necessary for servicing another component, you must compress the spring to remove it. Always use the proper kind of spring compressor for the job at hand, and always compress the spring the minimum amount needed to get it out of or into the mounting area.

To replace a coil spring, raise the vehicle and support it so the front wheels are off the ground. Remove the wheel. Disconnect the sway bar link and/or tie-rod end and any other components necessary to allow the lower control arm to be lowered.

Support the lower control arm. Install a suitable spring compressor to compress the spring. It may be necessary to remove the shock absorber for access. Separate the lower ball joint from the steering knuckle and lower the control arm enough to remove

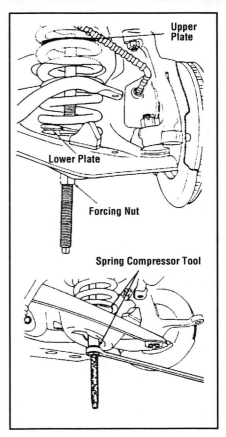

Compressing a coil spring prior to removal.
(Courtesy: Ford Motor Co.)

the coil spring. Inspect the spring insulators, if equipped, and replace as required.

Compress the new spring with the spring compressor. Install the spring insulators, if equipped, and position the coil spring. Make sure the spring is properly seated in the control arm and frame.

Raise the control arm and connect the lower ball joint. Remove the spring compressor and install the shock absorber, tie-rod end and sway bar link. Install the wheel and lower the vehicle.

LEAF SPRINGS

Leaf springs are used only on some truck front suspensions. Leaf springs are usually semi-elliptical in design. That means that they have a curved shape when viewed from the side. However, in recent years, improvements in spring

43

TRAINING FOR CERTIFICATION

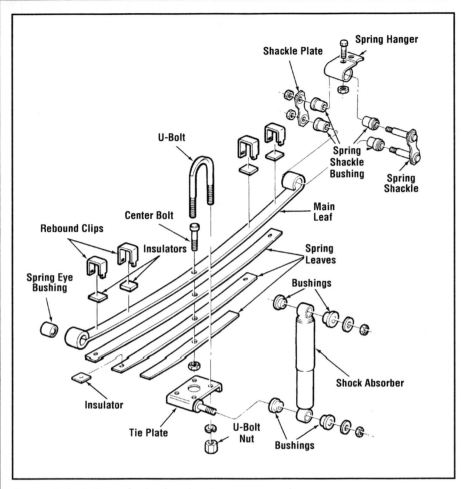

An exploded view of a leaf spring type suspension.

metal technology have reduced the amount of curve necessary. Newer springs may appear flat or almost flat with no reduction in rebound ability or load carrying capacity.

Generally, several spring leaves are used in combination, so that with increasing weight or deflection of the suspension, more spring leaves-and more spring force-come into play. Ordinarily, there are thin pads among the leaves to prevent them from rubbing and to prevent noise when the ends touch. Clips are used to prevent the leaves from pulling apart and clattering after extension and rebound.

Leaf springs are mounted at one end by a single bolt/bushing arrangement, which allows the spring to pivot with axle movement. At the other end, the spring is attached by a shackle setup. The shackle allows the spring to change in length as the vehicle encounters uneven road surfaces.

Leaf spring suspensions are relatively simple since the springs themselves serve as anchors for the axle, locating it fore-and-aft and side-to-side. Control arms and links are unnecessary. However, because of this design, leaf springs provide a harsher ride and are subject to sway because of their inherent inflexibility.

Inspect the leaf springs, hangers, shackles and bushings for wear and damage and replace parts as necessary. To replace a leaf spring, raise and support the vehicle so the front wheels are off the ground. Remove the wheel.

Support the axle and disconnect the shock absorber. Remove the nuts and U-bolts at the axle mount. Remove the shackle nuts/bolts and remove the hanger bolt. Remove the spring from the vehicle.

Install new bushings, as required. Position the spring in the vehicle and install the hanger bolt. Raise the other end of the spring and install the shackle bolts/nuts. Secure the axle to the spring with the U-bolts and nuts.

Connect the shock absorber and install the wheel. Lower the vehicle. Torque all fasteners with the vehicle at normal ride height.

TORSION BARS

A torsion bar is a length of spring steel that serves the same function as a coil spring. Torsion bars can be adjusted to correct ride height problems, therefore they are usually only replaced if they are damaged.

To replace a torsion bar, raise and support the vehicle at the frame so the front wheels are off the ground and the suspension is unloaded. Remove the wheel.

Remove the torsion bar adjusting bolt. Raise the lower control arm and remove the sway bar link, if necessary. Remove the torsion bar-to-control arm attaching bolts and the torsion bar anchor-to-frame bolts. Remove the torsion bar from the vehicle and remove the anchor from the torsion bar.

Check the replacement torsion bar for marks indicating on which side of the vehicle it should be installed. Most torsion bars are marked left and right and cannot be interchanged from side to side. Lubricate the hex end of the torsion bar and install it in the anchor.

Install the torsion bar in the vehicle and install the torsion bar-to-control arm bolts and the torsion bar anchor-to-frame bolts. Connect the sway bar link, as necessary. Install the adjusting bolt, turning it enough to load the torsion bar.

Install the wheel and lower the vehicle. Torque all fasteners to

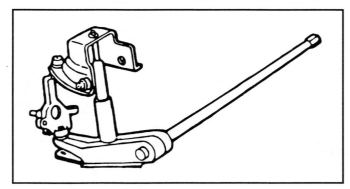

A longitudinally mounted torsion bar.

specification. Turn the adjusting bolt to set the vehicle ride height to specification. Check the front-wheel alignment.

SWAY BARS

The sway bar, also called a stabilizer bar or anti-roll bar, is a torsion-bar spring connecting the suspension on either side of the vehicle. When driving in a straight line along a flat road, the sway bar is doing nothing and is under no tension. When the body begins to roll to the side in a turn, however, the suspension at the outside wheel compresses and the suspension at the inside wheel extends. The sway bar that connects them twists to apply a counteracting force to hold the vehicle closer to level. A stiffer, that is, thicker, sway bar will apply more of this righting force and will make the vehicle corner 'flatter.' A softer, thinner sway bar will apply less righting force. This system is usually employed on vehicles with independent suspensions and light trucks with heavy front ends.

Sway bars are attached to the suspension arms and the body with rubber bushings. Sometimes, for reasons of clearance, there are also sway bar links between the bar and the suspension arm, and these links also have rubber bushings. While it is quite rare for a sway bar to break except from impact in an accident, it is not unusual for the sway bar bushings to wear out or sustain damage from oil leaking from the engine or power steering. Audible creaking or knocking sounds that occur only when a turn is commenced or ended are a typical symptom of bad sway bar bushings, and in most cases, visual inspection is fairly straightforward.

To remove a sway bar, raise the vehicle and support it so the wheels are off the ground and the suspension is unloaded. Remove the wheels. Remove the bolts or links attaching the sway bar to the control arms and remove the sway bar mounting bushing bracket-to-frame bolts. Remove the sway bar from the vehicle.

Position the sway bar to the frame and install, but do not tighten, the mount bushing bracket bolts. Install the sway bar end-to-control arm bolts or links, but do not tighten yet. Install the wheels and lower the vehicle.

With sway bar bushings, as with all suspension bushings, it is important to have the vehicle standing level on its wheels at normal ride height when the link or mount bushing bracket is tightened. Otherwise, the bushing will be twisted in normal use and will not last as long as it should. Also, some manufacturers specify that the link bolt be tightened only until the link bushings reach a certain stack height.

STRUTS

The most common type of strut used on vehicles today is the MacPherson strut. The MacPherson strut combines the shock absorber with the coil spring in a single strut, performing both suspension and suspension-damping functions. Some vehicles use a 'modified' strut suspension, which is identical to the MacPherson type in geometry, but it has the spring mounted between the frame and the lower (and only) control arm rather than on the strut. For some designs, this has engine compartment packaging advantages, and of course makes it simpler to replace the shock absorber.

The shock-absorbing working parts are held inside the lower strut tube by the collar nut and seal. Replacement of most MacPherson strut cartridges requires removal of the entire strut assembly, including the spring, from the vehicle. The unit is then

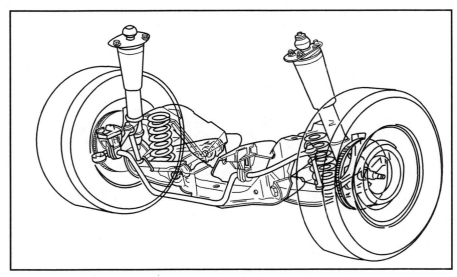

A typical end link type sway bar on a modified strut suspension.

45

TRAINING FOR CERTIFICATION

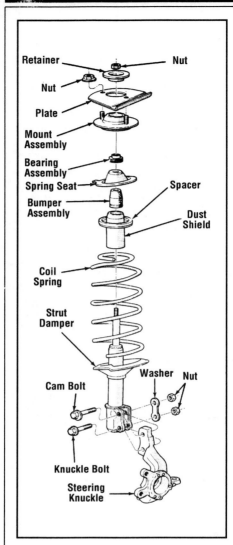

This is an exploded view of a typical MacPherson Strut used in a front suspension application. In this design, the strut is secured at the upper end to the vehicle's inner fender and at the lower end to the steering knuckle.

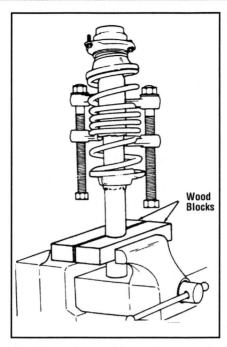

Always use a spring compressor to remove and install the coil spring. Compress the spring only as much as is necessary to remove and install the strut shaft nut.

the dampening effectiveness of new and old units may cause a stability-reducing roll after bumps.

To replace a strut cartridge, first mark the position of the strut in the strut tower, then loosen, but do not remove, the strut-to-strut tower nuts. Raise the vehicle and support it on the frame, so the wheels are off the ground and the suspension is unloaded. Remove the wheel and disconnect the brake hose from the strut.

If the strut is bolted to the steering knuckle, scribe alignment marks at the pinch bolt locations or strut-to-knuckle junction, remove the pinch bolts and separate the strut from the knuckle. It may be necessary to use a prybar to spread the pinch joint. If the strut is integral with the steering knuckle, remove the brake components and disconnect the tie-rod end from the steering arm, then disconnect the ball joint from the steering knuckle. Remove the strut-to-shock tower nuts and remove the strut from the vehicle.

Position the strut in a vise and install a spring compressor. Compress the spring and remove

placed in a spring-compressor fixture, and the upper bearing and then the collar nut and seal are removed. The small amount of heat transfer oil in the strut can be left in place, but should be replaced to make sure no water has entered the space and begun to rust the metal.

Most replacement cartridges include a new collar-nut and seal. While the strut is disassembled, you should check the upper bearing for any binding or roughness

and replace it if needed. Strut cartridges should always be replaced in pairs, particularly on light front-wheel drive vehicles; otherwise, the difference between

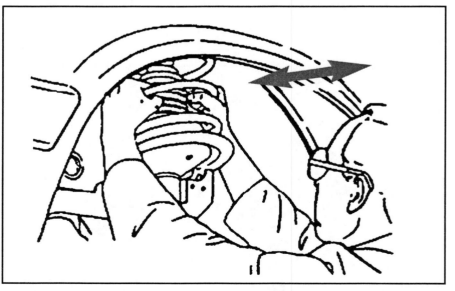

If the upper strut mount is in good condition, the strut piston shaft should not move freely when the strut is pushed in-and-out. Excessive free movement indicates the upper strut mount requires replacement.
(Courtesy: Tenneco Automotive)

the strut shaft nut. It may be necessary to hold the shaft while removing the nut. Remove the upper spring seat, coil spring and any insulators, jounce bumpers or dust shields.

Install the replacement strut in the vise and install the jounce bumper, dust shield, insulator, coil spring and upper spring seat. Install the spring compressor and compress the spring enough to install the strut shaft nut. Hold the shaft and torque the nut to specification.

Position the strut in the strut tower, aligning the marks made at removal. Install the strut-to-strut tower nuts finger-tight. If the strut is integral with the steering knuckle, connect the lower ball joint to the knuckle and the tie-rod end to the steering arm. Install the brake components.

If the strut is bolted to the steering knuckle, position the strut in the knuckle and install the pinch bolts. Align the scribe marks made at removal and torque the pinch bolts to specification.

Attach the brake hose to the strut and install the wheel. Lower the vehicle. Torque the strut-to-shock tower nuts to specification. Check the front-wheel alignment.

STRUT BEARING AND MOUNT

If the upper strut mount is defective, it may cause noise, steering binding, or allow the upper end of the strut to change position, affecting wheel alignment angles. A strut mount inspection should start with a road test checking for unusual noise, pulling or steering binding. If the road test results are satisfactory, bring the vehicle into the shop and listen for noises or binding as the steering wheels are turned from stop-to-stop with the vehicle's weight on the wheels. Noise or binding could indicate a defective bearing. Also inspect the rubber portions of the strut mount for cracks or separation of the rubber from the steel.

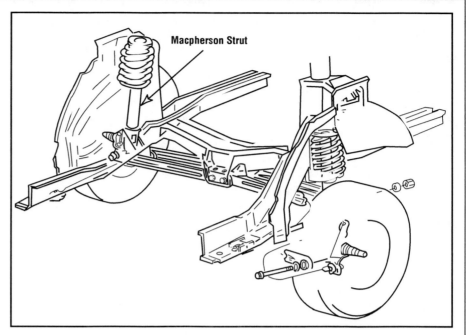

A rear suspension system with MacPherson struts.

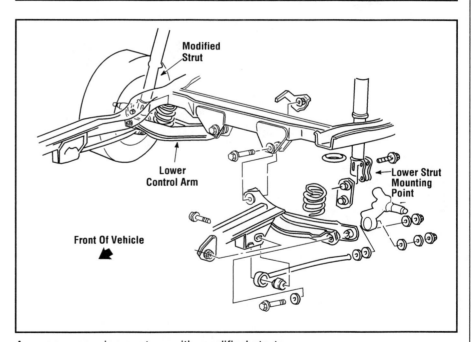

A rear suspension system with modified struts.

If the visual inspection reveals separation of the rubber portions from the steel strut mount, it should be replaced.

The mount can be inspected further with the vehicle on a lift. However, note the position of the strut piston shaft before raising the vehicle on the lift. As the vehicle is raised upwards on the lift, note any change in the position of the mount assembly. A slight downward movement is normal, but any side-to-side movement could indicate a defective mount.

Once the wheels are off the ground, grasp the coil spring as

TRAINING FOR CERTIFICATION

close to the upper strut mount as possible. Now, push in-and-out on the strut and spring while watching for movement of the upper end of the strut piston shaft. There should be no free movement. Excessive movement of the upper strut mount indicates that the mount requires replacement.

REAR SUSPENSIONS

There are many different rear suspension systems, but all designs can be categorized in one of two ways: independent and non-independent. An independent suspension allows wheels on the same axle to move up-and-down through the suspension travel separately from one another, and is a common form of front suspension. Up until a few decades ago, independent suspension was only used on the rear of exotic sports cars. However, as manufacturers began the switch to predominately manufacturing front-wheel drive cars, independent rear suspension became more common. It is easier to design and manufacture an independent suspension system when the added complexity of driving the wheels is eliminated.

Today, most vehicles, front-wheel drive and rear-wheel drive, have independent rear suspension. Non-independent, or solid axle suspension is now used mostly on trucks and some older design rear-wheel drive cars.

The types of independent rear suspensions in use range from those similar to designs commonly used in front suspensions, MacPherson strut, modified strut and double wishbone, to designs used almost exclusively on rear suspensions, such as the modified strut with transverse leaf spring. Non-independent rear suspensions use either coil or leaf springs. The leaf spring design is the simplest because the leaf springs also locate the rear axle. Coil spring non-independent suspensions use control arms or radius rods to locate the axle.

Rear Suspension System Diagnosis

Rear suspension systems are subject to most of the same problems detailed under Front Suspension System Diagnosis: noise, poor ride quality, abnormal tire wear, etc.

As with the front suspension, begin with a visual inspection, looking for obvious problems like leaking shock absorbers, worn bushings, broken springs and bent control arms. Check for loose wheel bearings and under-inflated and abnormally worn tires. If the vehicle leans to one side or sags in the rear, it may be due to broken or sacked spring(s).

Bounce each corner of the rear bumper to check the shock absorbers. If the vehicle rebounds more than twice, the shock absorbers are worn. Listen for noise as the suspension compresses and rebounds, which could be caused by worn bushings, struts or shocks. A wheezing or creaking sound may be caused by a shock or

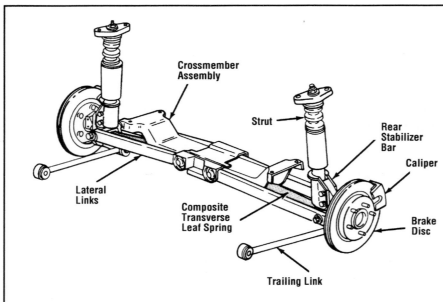

An independent rear suspension system with a fiberglass transverse leaf spring.

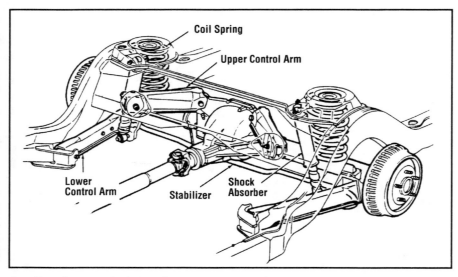

A coil spring non-independent rear suspension.

48

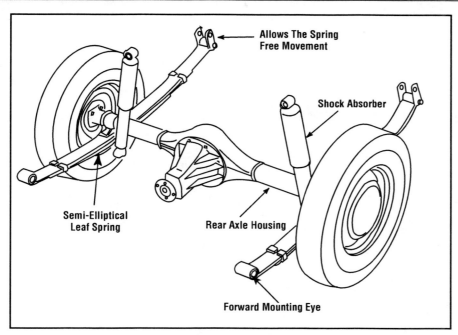

A leaf spring non-independent rear suspension.

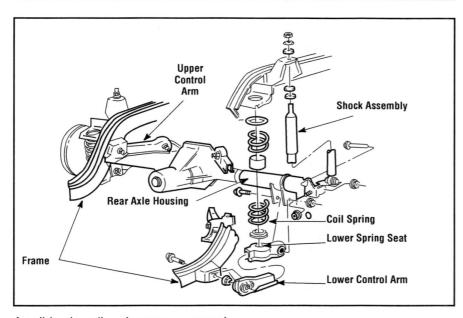

A solid axle coil spring rear suspension.

strut with dried out seals.

Listen for noise and observe the ride quality during a road test. Noise could be caused by worn bushings, loose shocks and worn spring shackles. A rough or harsh ride, excessive diving under braking, or excessive lean during cornering may be caused by worn struts or shock absorbers.

When checking suspension system component mounting and pivot points, make sure the suspension is unloaded and the area you are checking is not under tension.

Rear Suspension System Service

COIL SPRINGS

To remove a coil spring from an independent rear suspension it may be necessary to compress the spring with a spring compressor; it is usually not necessary on non-independent suspensions.

To remove the coil spring from the vehicle with independent suspension, raise the vehicle and support it so the rear wheels are off the ground. Remove the wheel. Support the lower control arm. Disconnect the shock absorber, sway bar link and any other components necessary to allow the control arm to be lowered.

If required, install a suitable spring compressor and compress the coil spring. Disconnect the lower control arm from the knuckle and lower the control arm. Remove the coil spring from the vehicle.

Installation is the reverse of removal. Torque all fasteners with the vehicle at normal ride height.

To remove the coil spring from a solid axle vehicle, raise the vehicle and support it on the frame so the rear wheels are off the ground. Support the axle with a jack, and then disconnect the shock absorber from the axle. Slowly lower the jack until the spring can be freed from the vehicle. Installation is the reverse of removal.

Coil springs should always be replaced in axle sets. When replacing coil springs, always inspect the insulators, if equipped, for damage and replace as necessary. During installation, make sure the spring is properly positioned in the spring seat.

CONTROL ARMS, TRAILING ARMS, LATERAL LINKS AND SWAY BARS

The great degree of rear suspension design variation prohibits a detailed discussion of the replacement procedures for these components. Inspect them for cracks, bending and worn bushings and service as required.

As a rule the suspension should be unloaded before removing an arm or link. Bushings are usually replaced using a press and suit-

49

TRAINING FOR CERTIFICATION

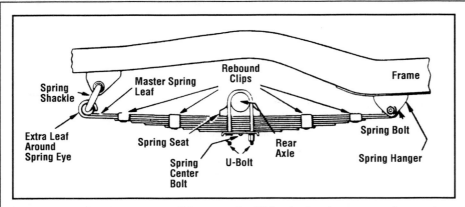

A typical leaf spring on a solid rear axle. This design is commonly used on light trucks now. As the axle moves up and down, the spring flexes and pivots on the shackle to accommodate its changing length.

able press fixtures. After arm or link installation, the fasteners should be torqued to specification with the vehicle at normal ride height, so the bushings will not be under strain.

LEAF SPRINGS

Inspect the leaf springs, hangers, shackles and bushings for wear and damage and replace parts as necessary. To replace a leaf spring, raise and support the vehicle so the rear wheels are off the ground. Remove the wheel.

Support the axle and disconnect the shock absorber. Remove the nuts and U-bolts at the axle mount. Remove the shackle nuts/bolts and remove the hanger bolt. Remove the spring from the vehicle.

Install new bushings, as required. Position the spring in the vehicle and install the hanger bolt. Raise the other end of the spring and install the shackle bolts/nuts. Secure the axle to the spring with the U-bolts and nuts.

Connect the shock absorber and install the wheel. Lower the vehicle. Torque all fasteners with the vehicle at normal ride height.

STRUTS

It may be necessary to remove interior or trunk panels to gain access to the strut tower mounting nuts. Mark the position of the strut in the strut tower, then loosen, but do not remove, the strut-to-strut tower nuts. Raise the vehicle and support it on the frame, so the rear wheels are off the ground and the suspension is unloaded. Remove the wheel and disconnect the brake hose from the strut.

If the strut is bolted to the knuckle or spindle, scribe alignment marks at the pinch bolt locations or strut-to-knuckle/spindle junction, remove the pinch bolts and separate the strut from the knuckle/spindle. It may be necessary to use a prybar to spread the pinch joint. If the strut is integral with the knuckle/spindle, remove the brake components, then disconnect the lower control arm from the knuckle/spindle. Remove the strut-to-strut tower nuts and remove the strut from the vehicle.

Position the strut in a vise and install a spring compressor. Compress the spring and remove the strut shaft nut. It may be necessary to hold the shaft while removing the nut. Remove the upper spring seat, coil spring and any insulators, jounce bumpers or dust shields.

Install the replacement strut in the vise and install the jounce bumper, dust shield, insulator, coil spring and upper spring seat. Install the spring compressor and compress the spring enough to install the strut shaft nut. Hold the shaft and torque the nut to specification.

Position the strut in the strut tower, aligning the marks made at removal. Install the strut-to-strut tower nuts finger-tight. If the strut is integral with the knuckle/spindle, connect the lower control arm and install the brake components.

If the strut is bolted to the knuckle/spindle, position the strut in the knuckle/spindle and install the pinch bolts. Align the scribe marks made at removal and torque the pinch bolts to specification.

Attach the brake hose to the strut and install the wheel. Lower the vehicle. Torque the strut-to-strut tower nuts to specification and reposition any interior or trunk panels that were removed. Check the wheel alignment.

NON-INDEPENDENT REAR AXLE

A non-independent, or solid, rear axle can be either the differential and axle housing assembly on a rear wheel drive vehicle or the beam axle on a front wheel drive vehicle. The designed position of the axle maintains the relationship of the rear wheels to the front wheels. Damage to the axle that causes misalignment can cause abnormal tire wear and vehicle tracking and handling problems.

Inspect the axle assembly for bending, warpage and misalignment. The damage may not be evident and a problem may only reveal itself through the aforementioned tire wear, and tracking and handling problems. In this case rear wheel alignment must be checked against specifications to find the cause and extent of the damage. Adjustment to bring alignment within specifications may not be possible or the amount of adjustment possible may not be sufficient, making axle replacement necessary.

BALL JOINTS AND TIE-ROD ENDS

The ball joints that connect the knuckle to the control arms, and the toe adjustment tie-rod end ball joint that attaches to the knuckle are both similarly constructed. They both have a ball that fits into a socket housing, with a tapered stud on the other end of the ball that fits into a tapered hole in the knuckle. Most ball joints are permanently lubricated, however some have grease fittings and can be lubricated during maintenance.

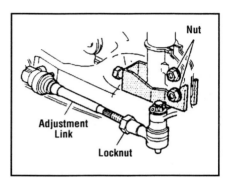

Rear tie-rod assembly.
(Courtesy: GM Corp.)

To inspect the control arm ball joint, raise and support the vehicle. Support the lower control arm and move the wheel up-and-down and in-and-out, checking for axial (up-and-down) and radial (side-to-side) play. To inspect the tie-rod end, grasp the tie-rod and move it up-and-down near the knuckle, then rock the tire back-and-forth. Disconnect the tie-rod end from the knuckle, double nut the stud and measure the effort required to turn the stud with a torque wrench. Compare any ball/socket play and turning effort with manufacturer's specifications.

Some ball joints have a wear indicator. If there is a shoulder into which the grease fitting threads, the ball joint is considered worn out once the shoulder recedes so far into the hole that its outer surface is level with the metal surrounding it. This type of joint must be replaced regardless of the amount of play detected in the ball socket. A ball joint also must be replaced if its dust cover is broken. Once grit can enter the joint, wear accelerates rapidly, and the joint will quickly fail.

Most ball joints fail due to wear in the ball and socket area, however, if the ballstud has broken, it is possible that the tapered hole has become distorted. You can check this by trying the new ball joint stud in the hole: if there is any free-play or if the new tapered stud can rock in the hole, the hole is rounded out. If this has occurred, the knuckle must be replaced.

To replace a control arm ball joint, raise and support the vehicle, then remove the wheel. Support the vehicle under the control arm or frame so the ball joint is not under tension. It may be necessary to use a spring compressor to compress the coil spring.

Remove the cotter pin and loosen the castle nut. Separate the tapered stud from the tapered hole using a puller designed for this purpose or by driving a ball joint separator, also known as a 'pickle fork', between the joint and knuckle. Keep in mind that using a ball joint separator will destroy the dust cover, making joint replacement mandatory. Unbolt the ball joint from the control arm.

Install the new ball joint to the control arm. Insert the ball stud into the knuckle and install the castle nut. Torque the castle nut to specification and install a new cotter pin. If the cotter pin hole does not align with the nut castellation, tighten the nut further just enough to install the cotter pin. Never loosen the nut to install the cotter pin. Install the grease fitting, if provided, and lubricate the ball joint. Install the wheel and lower the vehicle.

To replace a worn tie-rod end, loosen the locknut and disconnect the ball stud from the knuckle, using a puller or separator tool. As you unscrew the tie-rod end, count the number of turns required for removal. Thread the new tie-rod end on the same number of turns, so the toe setting will be close. Check and adjust the toe after the tie-rod end is installed. Torque the castle nut to specification and install a new cotter pin.

KNUCKLE/SPINDLE

The knuckle connects the upper and lower control arms or the strut and lower control arm and usually incorporates the spindle on which the wheel is mounted. A knuckle should be replaced if it is bent, cracked or if there is damage to the spindle or the tapered ball stud holes.

To remove the knuckle, raise the vehicle and support it so the rear wheels are off the ground. Remove the wheel and the brake components. Support the suspension so that tension is removed from the ball joint(s). Support the knuckle and separate it from the lower ball joint. Separate the knuckle from the upper ball joint or strut. On some strut vehicles, the strut-to-knuckle bolts also are used for camber adjustment. In this case, mark the position of the fasteners or the strut in the knuckle so that the camber setting will be the same after installation.

Install the knuckle and torque all fasteners to specification. Install the brake components and wheel, and lower the vehicle. Check the wheel alignment.

51

NOTES

NOTES

TRAINING FOR CERTIFICATION

RELATED SUSPENSION AND STEERING SERVICE

SHOCK ABSORBERS

The function of shock absorbers is not to 'absorb shocks' (that's what the springs do), but to dampen the oscillation of the suspension, that is, to stop the bouncing up-and-down that would otherwise occur after a spring deflection. As such, the shock absorbers do far more of their work as they are extending rather than while they are compressed, because the springs will resist compression during 'jounce' (compression), but not during 'rebound' (extension).

Shock absorbers work by pushing and pulling a vented piston through a gas- or oil-filled chamber. Fixed- and variable-diameter orifices control the resistance with which the absorber resists rapid extension or compression.

Since they are sealed units, there are no repairs possible on shock absorbers beyond replacement when they have worn out or failed. With time, enough air and moisture enters the system through the shaft seal to contaminate the gas or oil. As the piston is driven back and forth, the moisture can vaporize under vacuum and compromise the dampening properties of the shock absorber.

Sometimes physical damage or wear allows the shock absorber to leak. In such cases, overheating sometimes damages the remaining oil. Nitrogen gas-filled shock absorbers are designed to delay this condition by displacing air and moisture with a positive pressure, dry nitrogen gas charge. Most shock absorbers are cooled merely by exposure to air.

Shock absorber inspection is covered in the suspension systems diagnosis sections of this study guide. To replace a worn shock absorber, raise and support the vehicle. If removing the rear shocks from a vehicle with solid axle suspension, support the rear

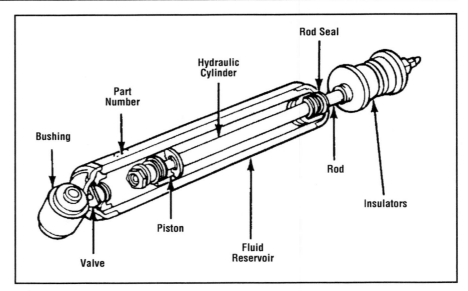

Cross-sectional view of a typical shock absorber. *(Courtesy: Ford Motor Co.)*

axle. Remove the upper and lower shock fasteners and remove the shock absorber.

Before installing the new shock, it must be purged of air. With the shock absorber right side up, extend it fully, then turn it upside down and fully compress it. Repeat the procedure three more times to make sure any trapped air has been expelled.

WHEEL BEARINGS

Every bearing is essentially a set of wheels within wheels - a set of small, fast spinning wheels between larger, slower-turning ones. The metal in bearings must be incredibly hard to withstand the constant high pressure of the weight of the vehicle and its maximum impact load in a hard-struck rut concentrated onto very small areas: the actual contact areas between the balls or rollers and the races.

The average life of a wheel bearing is predicted to be in the 150,000-mile range, but half will fail before that figure. Visual inspection of the bearing is useful only for determining when to discard them. If there is almost any visible wear to the mirror-like working surfaces, you can be sure that 99 percent of the life of that bearing is past.

In recent years, wheel-bearing replacement has become very simple because the bearings

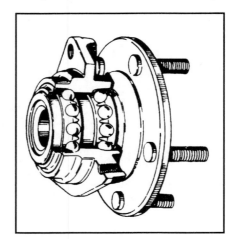

Wheel bearings for FWD vehicles often come as complete units with a mounted flange and hub. The drive axle extends through the center of the bearing and is secured by a precisely torqued axle nut.

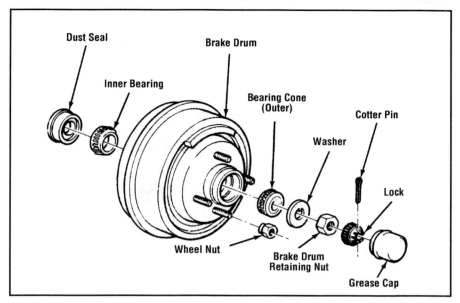

This exploded view shows all the replaceable parts in a typical, replaceable tapered roller bearing assembly. All parts are serviceable and don't forget to replace the dust seals whenever the bearings are serviced or replaced.

themselves can't be replaced. Newer vehicles come equipped with integral bearing/hub assemblies. These unitized components are used on both front and rear axles, regardless of whether the vehicle is front- or rear-wheel drive. The problem with integral assemblies is that the bearings can't be visually inspected. Inspection is limited to feel and hearing. Inspections consist chiefly of turning the wheels; listening/feeling for any grinding or crunching sounds, even very subtle ones; and moving the wheel axially to see how much free-play there is in the bearings. At one time, with rear-wheel drive vehicles, as much as 1/16-in. free-play was considered an allowable amount in the front wheel bearings, but on modern vehicles that figure has been reduced. Check with the manufacturer's specifications to be sure, but generally any free-play is too much.

Two kinds of bearings are used for wheels—ball bearings and tapered roller bearings. Generally, tapered rollers are somewhat more forgiving and tolerant of preset torque. The adjustment is often set just finger-tight and locked in that position by cotter pins or tabs and lock nuts. Ball bearings must be precisely tightened with a torque wrench to exactly the proper specification. Most ball bearings are factory-lubricated and sealed for life; they can't be repacked with grease. On some front-wheel drive vehicles, removal of the front-wheel bearings destroys them, so new ones are necessary each time. With others, a number of shims and/or spacers are used between the inner and outer bearings. These pieces are usually discarded and must be replaced with new ones to properly set bearing preload. With integral bearing/hub assemblies, no bearing preload or other settings are required. The assembly just bolts into place.

If a bearing has failed, carefully inspect the working surfaces of the spindle on which it worked. You are looking for two flaws: abrasion damage and discoloration, particularly heat discoloration like metal bluing.

Abrasion damage can be the beginning of a crack that would allow the spindle to fail catastrophically. Metal bluing indicates a change of the internal crystalline structure of the metal: it has become hardened and is no longer strong enough to support the kinds of forces transmitted through a wheel.

In each case, replacement usually involves replacing the spindle or spindle/knuckle assembly.

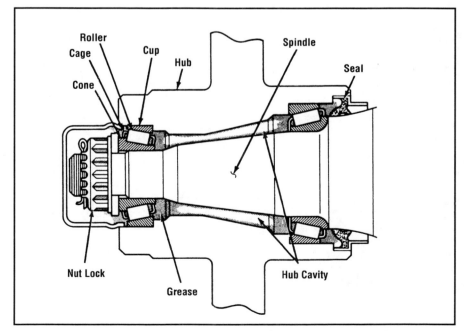

A tapered roller bearing assembly on a spindle. Inner bearings are usually larger than outer bearings due to weight distribution and loading.

55

TRAINING FOR CERTIFICATION

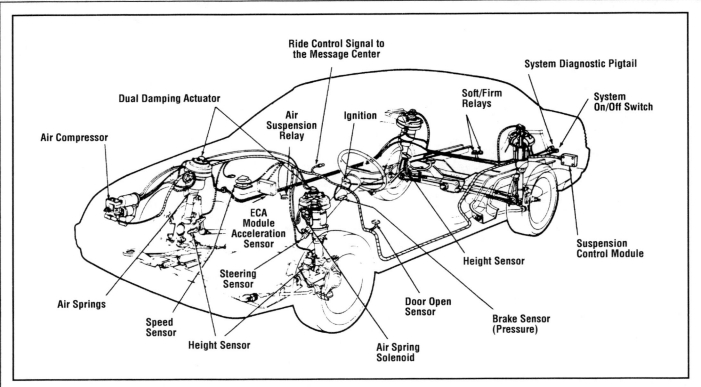

Typical electronically controlled air suspension system. *(Courtesy: Ford Motor Co.)*

ELECTRONICALLY CONTROLLED SUSPENSION SYSTEMS

There are many types of electronically controlled suspension systems, ranging from relatively simple electronic leveling systems that maintain a vehicle's ride height regardless of load, to sophisticated adaptive systems that control vehicle ride and handling by continuously altering ride height and shock damping.

There are pneumatic and hydraulic systems, and some systems that use both. Pneumatic systems use air shocks instead of hydraulic shocks or air springs in place of or in addition to the usual torsion bars, or coil or leaf springs. The air springs are used to maintain the vehicle ride height or to change the spring rate for particular driving conditions. Air shocks are used to raise or lower the vehicle to maintain ride height.

Hydraulic systems use actuators that change the valving in the shocks or struts. A small electric motor on the shock or strut rotates a rod, which alters the size of the metering orifices in the shock or strut.

All systems use sensors to provide input data to a computer, which then, depending on the system, commands an on-board air compressor to inflate or deflate an air spring or air shock, or operates the actuator motor on the shock or strut to change the damping. The type and number of sensors used varies according to system.

Height sensors, suspension levers connected to potentiome-

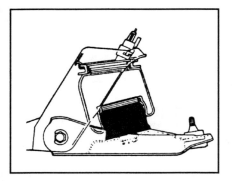

Here an air spring is used in place of a coil spring.

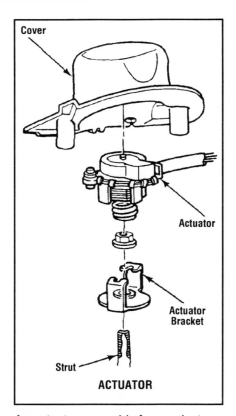

An actuator assembly for an electronically controlled MacPherson strut. *(Courtesy: Ford Motor Co.)*

56

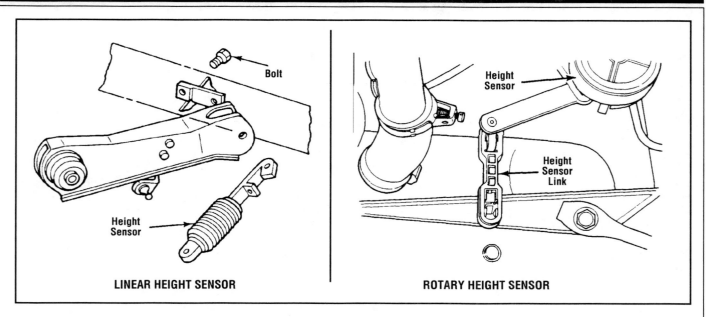

Two types of height sensors. *(Courtesy: Ford Motor Co.)*

ters, signal ride height data to the computer. In some systems they also signal road undulation information to the computer, such as would be encountered when crossing railroad tracks. The computer can in turn firm up the suspension to prevent bottoming out.

The steering wheel rotation sensor notes steering wheel position and measures how fast the steering wheel is being rotated. Rapid steering wheel rotation or a hard turn at high speed would signal the computer that a firmer suspension was needed, such as would be required when performing an evasive maneuver.

The vehicle speed sensor signal is used by the computer to determine when a firmer suspension is needed. The suspension setting is automatically changed when the vehicle reaches a certain speed and then returns to a softer set-

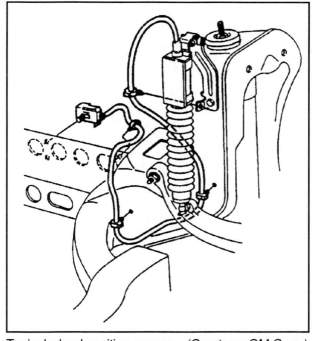

Typical wheel position sensor. *(Courtesy: GM Corp.)*

ting when the vehicle slows. The suspension may also be firmed if vehicle speed along with a particular steering angle combine to exceed a certain lateral acceleration figure.

Wheel position sensors are mounted at each corner of the vehicle, between a control arm and the body. The input from the these sensors provides the computer with relative wheel to body position and velocity. Rapid sensor movement with a large amount of sensor travel would indicate that the wheel was moving at high speed over a rough surface, which would require a firmer suspension setting.

Some of the more advanced electronically controlled suspension systems also use input from vacuum and throttle position sensors to firm the suspension during acceleration. Other systems use yaw sensors to sense body roll when cornering, and G-force sensors to sense sudden braking or acceleration.

System Diagnosis

Always consult the service manual for the particular vehicle in question for the proper diagnostic procedures. Make sure the conventional portion of the suspension: the ball joints, bushings, etc., are in good condition. Visually inspect the electronic components, and check the wiring for damage and secure connections.

On pneumatic systems, check for leaks in the lines or fittings. The best way to discover these leaks is by brushing soapy water

TRAINING FOR CERTIFICATION

on the suspected components. The next most common problem comes from moisture in the lines, an inherent problem for any air compressor system. Most have either an air drier or some device to vent the moisture. If these measures fail, there can be a problem with water and internal rust on some systems, and with ice in winter on all of them.

A failure in an electronically controlled suspension system will usually set a Diagnostic Trouble Code (DTC), which will be stored in the memory of the system computer and in most cases, illuminate a light on the instrument panel. The light alerts the driver that there is a problem in the system and may say 'Check Ride Control' or 'Service Ride Control'.

DTCs can be retrieved on some systems by pressing a combination of buttons on the heating and air conditioning controls or on the vehicle message center. The codes are then displayed on the instrument panel. Codes can also be retrieved using a scan tool connected to the vehicle Diagnostic Link Connector (DLC).

Once you have the DTCs that require diagnosis, refer to the appropriate service information to identify the circuits that the DTCs represent. The diagnostic charts will describe the circuit and the fault that the code represents and contain troubleshooting procedures and tests that must be performed, to determine the cause of the malfunction. These tests usually describe various voltage and resistance measurements using a Digital Multimeter (DMM), however some systems specify tests that must be performed with a bi-directional scan tool.

Service Precautions

On most vehicles with air suspension, the air suspension switch must be turned off before the vehicle is raised on a hoist or towed.

Many manufacturers recommend that vehicles with air suspension systems be raised on the frame only and not on the suspension.

Always be sure to exhaust the air from an air spring before removing it or any components supporting the air spring.

SUBFRAMES AND CROSSMEMBERS

Many vehicles have the drivetrain, suspension and/or steering components mounted to removable subframes or crossmembers. It stands to reason then that if the subframe or crossmember is damaged or not properly positioned and secured to the vehicle, the aforementioned systems will be affected.

As was mentioned in Steering Systems Diagnosis, an incorrectly positioned subframe can cause binding in the steering column. A loose or incorrectly positioned subframe could also cause handling and alignment problems.

Inspect the subframe or vehicle crossmember and the mounting bushings and brackets for wear and damage. If a crossmember or subframe is lowered for drivetrain or other system service, it must be correctly reinstalled. Usually there are alignment holes in the body and subframe, which should align if the subframe is installed correctly. Some manufacturers specify that the fasteners be installed and tightened in a certain order. Refer to the vehicle service manual for instructions.

ELECTRONICALLY CONTROLLED STEERING SYSTEMS

The knock against power steering systems has been that they don't give the driver enough road feel, especially at medium to high speeds where a feeling of increased control is desirable. A fully mechanical power steering system can never be more than a compromise: if there is high power assist, there is reduced road feel; if they provide good road feel there is reduced power assist, which hinders low speed driving and parking maneuvers.

The addition of electronic controls to the power steering system enables the system to reduce steering effort at low speeds and increase steering feel and control at high speeds, by varying assist according to vehicle speed. The steering system control module receives inputs from the vehicle speed sensor and the steering wheel rotation sensor.

Since less power assist is needed at higher speeds, the speed sensor data is used by the control module to reduce power assist as vehicle speed increases. The steering wheel rotation sensor provides input to the control module regarding steering wheel position and steering wheel turning rate. The steering wheel position data allows the control module to calculate lateral acceleration, or the speed the vehicle is moving during a turn. The steering wheel turning rate indicates when the steering wheel is being turned rapidly, as would be the case during a high speed evasive maneuver. The control module, in turn, increases power assist when there is high lateral acceleration or rapid steering wheel rotation.

The control module usually varies power assist by varying fluid pressure to the boost piston in the steering gear, using either

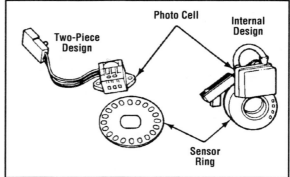

Typical steering rotation sensor.
(Courtesy: Ford Motor Co.)

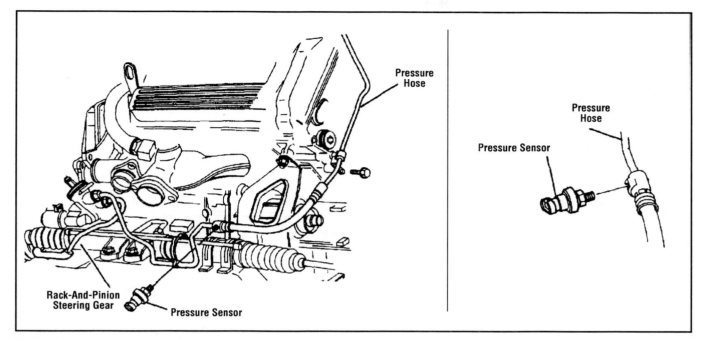

Typical power steering pressure sensor mounting location.

a pulse-width modulated solenoid valve or stepper motor operating a variable orifice valve. The exception to this is General Motors' Magnasteer, which varies power assist without varying fluid pressure. Magnasteer uses electromagnets mounted in the steering column to increase or decrease the torque needed to turn the steering wheel.

System Diagnosis

If a variable assist power steering system fails, a change in steering effort will usually be noticed, such as having full power assist at higher speeds or that the level of assist is not consistent. Most systems are designed to provide full power assist in the event of a system malfunction.

Begin diagnosis by making sure the mechanical part of the power steering system is in good condition. Make sure there are no fluid leaks or damaged components. Visually inspect the electronic components, and check the wiring for damage and secure connections.

Always consult the service manual for the particular vehicle in question for the proper diagnostic procedures. A failure in the system may set a Diagnostic Trouble Code (DTC), which will be stored in the memory of the system computer and in most cases, illuminate a light on the instrument panel. DTCs can be retrieved using a scan tool connected to the vehicle Diagnostic Link Connector (DLC).

Consult the appropriate service information to interpret codes. The service information will contain wiring diagrams, electrical values and troubleshooting procedures and tests that must be performed, to determine the cause of the malfunction. These tests usually describe various voltage and resistance measurements using a DMM.

POWER STEERING IDLE SPEED COMPENSATION SYSTEM

The power steering pump, like any accessory driven off the engine's crankshaft, exerts a load on the engine. This load is especially high when the hydraulic pressure in the system is high, such as occurs during parking maneuvers when the steering wheel is turned against the stops while the vehicle is stationary.

Most newer vehicles are equipped with a sensor that informs the Powertrain Control Module (PCM) of high power steering pressure, causing the PCM to in turn adjust the Idle Air Control (IAC) valve or Idle Speed Control (ISC) motor to raise the engine idle speed. Raising the idle speed prevents the engine from stalling during periods of increased load.

A failure in the system will usually result in high idle speed or stalling and may set a DTC. A scan tool can be used to check the status of the system. To test the pressure sensor, jumper the connector terminals and look for a change in engine RPM.

NOTES

NOTES

TRAINING FOR CERTIFICATION

WHEEL ALIGNMENT DIAGNOSIS, ADJUSTMENT AND REPAIR

Wheel alignment is the system of angles designed into the vehicle's suspension that when properly adjusted, allow the tires to roll freely and make the vehicle handle in a safe, predictable manner. Some of these angles, caster, camber and toe, can be adjusted when performing a wheel alignment, while others, such as toe-out on turns and steering axis inclination, can only be changed by replacing parts or straightening the frame.

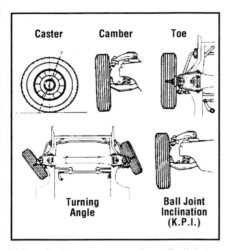

Wheel alignment angles. Ball joint inclination, also known as steering axis inclination, and turning angle, also known as toe-out on turns, are not adjustable during an alignment. *(Courtesy: Ford Motor Co.)*

DIAGNOSIS AND INSPECTION

A wheel alignment is performed to cure a vehicle handling problem, stop abnormal tire wear, or because suspension or steering components have been replaced. Of course, the cause of poor handling and tire wear may not be caused directly by incorrect wheel alignment, and may be due to worn suspension or steering components. This can be determined during the vehicle inspection.

When a vehicle is presented for an alignment, first take it for a test drive. If the vehicle owner was reporting a problem, the owner should drive the vehicle. While the owner's diagnosis may not necessarily be correct, the person who drives a vehicle every day is the one most likely to notice a problem with steering or handling.

During the road test, pay particular attention to steering pull, wander, unusual noises either straight ahead or in a turn, recovery from turns, and the stability of the vehicle in each of these maneuvers. Note if the vehicle owner employs driving techniques (such as braking in a turn) that lead to handling instability.

Try to do the road test on a familiar route, one with steep and shallow turns, hills, parking operations and highway speeds. In this way you can quickly be alerted to an individual vehicle's steering and suspension flaws. Drive the vehicle yourself to obtain first-hand information about the vehicle's handling characteristics.

After the test drive, inspect the vehicle for worn or damaged components that could cause the handling problems experienced during the test drive.

Make sure the tires are all the same size and are the correct ones for the vehicle. Inspect the tires for proper inflation and unusual wear. Underinflated tires will show wear on the edge or shoulder of the tire. Overinflation wears out the center of the tread. Incorrect toe will cause the edges of the tread to feather, while incorrect camber will cause the tire to wear one side of the tread. Sometimes tires can cause the vehicle to pull to one side. If this is the complaint, try switching the front tires from side to side. If the vehicle now pulls to the other direction, the cause is the tires.

Raise the vehicle and inspect the steering and suspension components. Look for bent or broken components and evidence of collision damage. Some vehicles have bolt-on subframes or cradles on which the drivetrain, suspension and/or steering is mounted. If the subframe or mounting bushings are damaged in a collision or the subframe is incorrectly installed, wheel alignment could be affected. Usually there are alignment holes in the body and subframe, which should align if the subframe is installed correctly. In addition, some manufacturers specify a particular fastener installation and tightening procedure when installing the subframe. Refer to the vehicle service manual for instructions.

Check the wheel bearing adjustment and check the ball joints for excessive play. Both can cause camber to be incorrect, which in turn can cause pulling and tire wear.

Check the control arm bushings and strut rod bushings for looseness and wear. These can affect caster, causing pulling or steering problems. Loose or worn control arm bushings can also affect camber.

On MacPherson strut equipped vehicles, check the upper strut mount for looseness and wear as this could affect caster and camber.

Inspect the steering linkage for excessive play, which could cause shimmy and incorrect toe.

Lower the vehicle and check for worn shocks or struts by bouncing the vehicle at each corner. Check the vehicle ride height. Consult the vehicle service manual or the alignment machine handbook to determine where to measure the vehicle's at-rest ride height. On some vehicles, it is measured from the pavement surface to the rocker panels; on

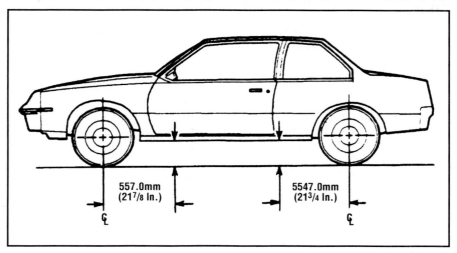

On this vehicle, the ride height is the measurement of the distance between the rocker panel and the ground.

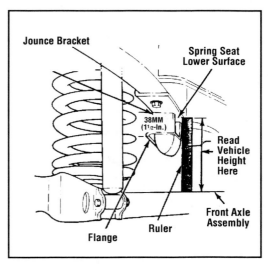

A ride height measurement taken between suspension components.
(Courtesy: Ford Motor Co.)

others, it is measured more precisely by angles or in distances between suspension components. Except in the case of vehicles with adjustable torsion bar suspensions, ride height is corrected by replacing springs or shimming them with special spacers that fit in the spring sockets.

In addition to the preceding, you should also be aware of the following steering problems:

'Torque steer' is the tendency of many front-wheel drive vehicles to turn somewhat from the desired direction when accelerating, especially in a curve, or when decelerating under engine vacuum retard in a curve. This is a much-disputed characteristic of some of these designs, most probably related to front halfshafts of unequal length. There is no repair procedure to correct this situation.

'Bump steer' is what happens when the vehicle wants to dart one way or another just as the suspension goes through jounce-and-rebound. This is caused by a change of toe setting as the suspension works, most commonly because the springs have sagged or because the steering rack or idler arm has somehow gotten out of parallel with the horizontal line of the front axle. Inspect for this problem by working the suspension with the vehicle on an alignment rack, wheel plates free. It is not uncommon for one side of the steering to show this change of toe, but not the other. Note that bad tires can mimic the symptoms of bump steer, including bad tires on the rear wheels.

"Memory steer' is the characteristic of some vehicles to want to go in a specific direction other than straight ahead. The most frequent causes of this are binding upper bearings on strut systems. This can also lead to memory steer that changes after the binding bearing shifts to a new position. In this situation, the spring is twisting where the bearing should be turning, but unlike the bearing, the spring has a position to which it wants to return. Replacement of the affected bearing is the solution.

Another common form of memory steer occurs in vehicles with RBS tie-rod ends. This may occur if a technician who previously worked on the vehicle did not straighten the wheels before tightening the bolt on the tapered stud, or if the tapered stud turned while being tightened. If the rubber and the joint are still in good condition, that is, if the situation has not existed for too long, the joint can be separated and reconnected in the proper way. Replacing only one RBS tie-rod end may also cause this condition. In this case, a memory steer condition will exist towards the side with the newer RBS tie-rod. It is recommended that this type of tie-rod be replaced in pairs to avoid this condition.

All worn or damaged parts found during the inspection procedure must be replaced before the vehicle can be aligned.

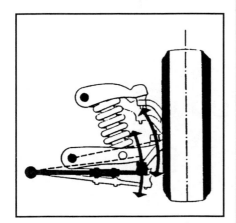

When the suspension and steering systems are intact, the tie-rod end and the ball joints move concentrically. If damage or wear has thrown them out of this arrangement, bump steer can result.

63

TRAINING FOR CERTIFICATION

DESCRIPTION OF ALIGNMENT ANGLES

Camber

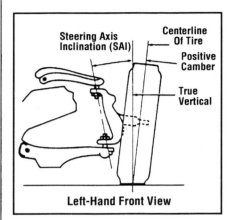

Camber is the angle as seen from the front of the vehicle between a wheel's centerline and a true vertical. Positive camber leans out from the vehicle at the top and negative camber leans in at the top. A vehicle will tend to steer or pull towards the side with the most positive camber. A general rule of thumb is that camber should not vary by more than 0.5 degrees from one side to the other.

Camber is the angle between the vertical centerline of a wheel/tire and the true vertical as seen from the front. Most vehicles have a slightly positive (leaning slightly out at the top) or neutral camber on the front wheels. On vehicles with an adjustable rear camber, it is usually set at neutral or slightly negative (leaning slightly in at the top). More positive camber on the front wheels will cause a pull to that side. Excessive positive camber will cause accelerated wear of the tire on the outside edge and excessive negative camber will cause accelerated wear on the inside edge of the tire.

Caster

Caster is the angle formed between a true vertical line and a line through the rotational center of a vehicle's steering axis as seen from the side of the vehicle. Caster provides the inclination of the vehicle to travel in a straight line forward. It also helps improve steering wheel returnability after turns. Excessive caster makes it hard to steer to either side and can increase road shock and shimmy. Insufficient caster will result in a vague, wandering steering feel. On a few very heavy vehicles, caster is set almost neutral or even slightly negative to overcome the problems of turning against heavy resistance, but on most vehicles, caster is set positive. Many high performance vehicles are engineered with significant positive caster, as much as 12 degrees, to retain directional stability at high speeds. If no specification is provided, it is a general rule-of-thumb, that caster should not vary by more than 0.5 degrees from one side to the other.

On most vehicles, caster will incline the steering to either the right or left bump stop in reverse. When moving forward, a vehicle will tend to steer towards the side with the least amount of caster.

Toe

Toe is defined as the difference in distance measured across the front and across the rear of the tires. Toe can also be defined as a comparison of horizontal lines drawn through both wheels on the same axle. A wheel is said to be 'toed-in' if its line of forward direction intersects the extended centerline of the vehicle; a wheel is 'toed-out' if its line of forward direction and the vehicle centerline are angled apart. In both cases, this angle is very small.

On rear-wheel drive vehicles, steering is generally toed-in, so that under load and under braking, the steering will become more nearly neutral. Front-wheel drive vehicles are often slightly toed-out, so they will come to neutral under

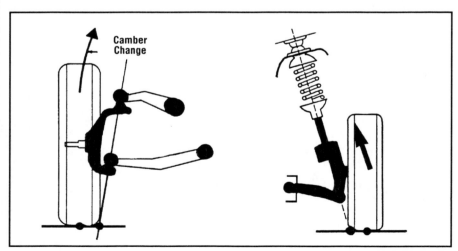

On Short/Long Arm (SLA) type suspensions (left), there is a positive change in camber as the suspension moves up and a negative change as the suspension rebounds or moves downward. On strut type suspensions (right), the change is opposite. Camber moves in the negative direction as the suspension is jounced or moved up and camber moves in the positive direction as the suspensions rebounds.

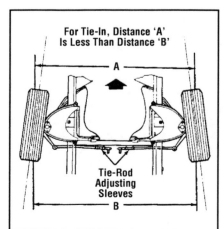

Toe Measuring points.
(Courtesy: DaimlerChrysler Corp.)

64

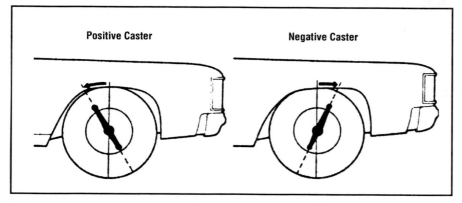

Caster is the forward or rearward tilt of the steering axis in reference to a vertical line. The caster angle is always measured and displayed in degrees. Caster is positive when the top of the steering axis is tilted rearward and it is negative if the top of the steering axis is tilted forward.

power. With rack-and-pinion steering, toe is set closer to absolute neutral on most vehicles, whether front- or rear-wheel drive.

Toe is not an angle that affects handling directly (i.e., it will not cause a drift or pull). However, it is the most significant tire-wearing angle. Excessive toe-in or toe-out will cause rapidly accelerated tire wear, as the tire slides sideways while the vehicle moves forward.

Individual toe is the difference in distance of the front and rear of one tire in reference to a centerline extended from the vehicle.

Steering Axis Inclination (SAI)

SAI is the angle formed between a true vertical line and an imaginary line through the centerline of the ball joints, or on strut suspensions, through the centerline of the ball joint and the upper bearing, as seen from the front of the vehicle. When solid axles with kingpins were more common, this was referred to as kingpin inclination. On these vehicles, the imaginary line is drawn along the axis of the kingpin.

Notice that this measurement uses the exact same reference line as when measuring caster, but seen at an angle 90 degrees away from the front of the vehicle instead of from the side.

If the SAI is wrong on an otherwise correctly aligned vehicle, the spindle or the steering knuckle is bent. On most vehicles other than some four-wheel drive trucks, heavy-duty pick-ups and some vans, this is the same part.

Included Angle

The included angle is the angle formed between an imaginary line drawn through the centerline of the ball joints (or on strut suspensions, between the centerlines of the ball joint and the upper bearing) as seen from the front, and the centerline of the tire/wheel. The included angle is equal to SAI plus camber. A large difference in side-to-side included angle indicates that a bent component most likely exists between the pivot points of the steering axis.

Always measure SAI and included angle with the wheels straight ahead. This will prevent camber roll from influencing the measurement and falsifying the readings.

Scrub Radius

The distance between the point at which the tire's vertical centerpoint intersects the road, and the SAI intersects the road is the scrub radius. This is a linear dimension rather than an angle. It is so called because it is the distance the tire must slide or 'scrub' along the ground if it is turned when the vehicle is standing still. Designers engineer this dimension to affect the handling qualities of the vehicle under conditions of brake failure or tire blowout. While actual numbers are not always provided in alignment tables, an inaccurate scrub radius indicates the same damaged components as incorrect SAI.

Toe-Out On Turns

When a vehicle makes a turn, the inside wheel drives around a

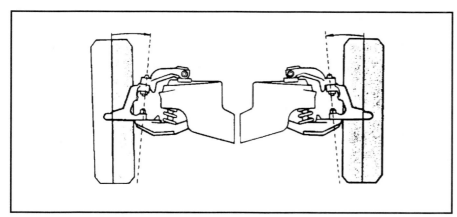

Steering axis inclination (SAI), also known as ball joint or kingpin inclination, is the angle formed by a line drawn through the upper and lower pivot points of the spindle and true vertical as viewed from the front. SAI permits the use of smaller camber angles by helping maintain the vehicle load inboard on the spindle. Greater load inboard on the spindle enhances road isolation. If SAI is incorrect and camber is within specification, something in the suspension is most likely bent.
(Courtesy: Ford Motor Co.)

TRAINING FOR CERTIFICATION

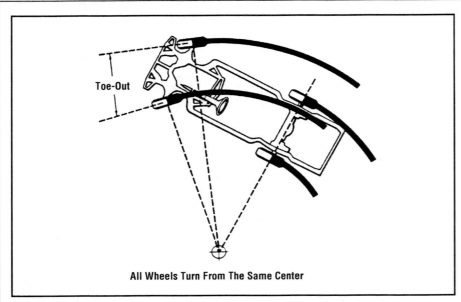

All Wheels Turn From The Same Center

When a vehicle makes a turn, the inside wheel makes a smaller circle than the outside wheel. The steering arm is designed to account for this and turn the inside wheel more steeply.

smaller curve than the outside wheel. If the steering geometry does not compensate for this, the inside wheel, being less heavily loaded, will slide sideways and lose tread life. Consequently, the angle of the steering arm on the steering knuckle is engineered to splay the front wheels slightly apart from one another in a turn. This is true whether the straight-ahead toe specification is slightly toed-in or slightly toed-out. All vehicles are toe-out on turns. This toe-out on turns should be equal at equal turning angles from side-to-side if it is a symmetrical design. However, some systems are not symmetrical, this means that the turning angle will be unequal at the same turning angle from side-to-side. Always consult the manufacturer's specifications for that vehicle before making a judgment.

If the turning angle is not in compliance with manufacturer's specifications, the steering arm should be inspected. As a general rule-of-thumb, turning angle problems will not be noticeable unless the turning angle exceeds the manufacturer's specification by more than 1.5 degrees.

Some vehicles, notably trucks, have adjustable steering stops at the extremes of steering angle. Ordinarily, this dimension is not found in alignment tables and it is seldom checked in practice, but it may be a worthwhile test if a tire strikes a fender, the frame or a suspension component. In general, steering stops do not go out of adjustment unless there is damage or inaccurate work. Sometimes, the stops should be set slightly differently if larger wheels and tires are installed.

Thrust Angle

Thrust angle is the average toe of the rear wheels. This is true whether the vehicle is front- or rear-wheel drive. The term was chosen when most vehicles were rear wheel drive, and it is not fully accurate anymore, but has survived anyway. The thrust angle determines the angle of the centerline of the vehicle with the road when driving in a straight line. Obviously, on all vehicles the ideal thrust angle is zero. On some vehicles, particularly those with rear-wheel drive and solid rear axles, making small changes in thrust angle is practically impossible. Large inaccuracies occur only when the axle slips on the leaf springs or, with rear coil springs, an axle control arm or

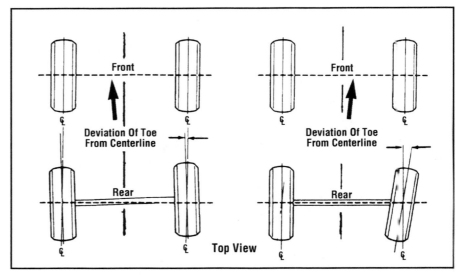

Thrust angle is the bisector of the rear total toe, or the direction the rear wheels are pointed. This definition applies regardless of whether the vehicle is front or rear wheel drive. Incorrect vehicle thrust line makes the vehicle dog track and requires counter steering of the front wheels in order to keep the vehicle traveling in a straight line. A four-wheel alignment is preferred on a vehicle with adjustable rear suspension. This allows for all four wheels to be aligned to the vehicle's geometric centerline. However, thrust line alignments are suitable for any vehicle equipped with a non-adjustable rear suspension.

bushing is damaged.

Try to set the thrust angle identically with the vehicle centerline. If that is impossible or prohibitively expensive, align the front wheels to the thrust angle. This means the first step in alignment after the inspections is to determine the thrust angle. Any work done without knowing the thrust angle may have to be undone as soon as it is tested. This is just as true for older designs with virtually fixed rear axles as for newer, all-independent suspensions.

Setback

Usually a vehicle's wheelbase (the distance from front to rear wheel center) is the same on both sides. On a few vehicles, such as light trucks with Twin I-Beam front suspension, the front wheel is slightly farther back on one side. Ordinarily, however, if you find setback on a vehicle, it means there has been frame or unibody damage that has shifted the mounting points of the suspension components. With front-wheel drive vehicles, keep in mind that damage to the rear suspension mounting areas can have a similar effect.

Setback is the angle formed by the geometric centerline and a line drawn perpendicular to the front axle. Positive setback indicates that the right front wheel is positioned further rearward than the left front wheel. Hence, the wheelbase will be shorter on the right as opposed to the left. Likewise, a negative setback indicates that the left front wheel is further rearward than the right front wheel.

Small amounts of setback can be accommodated in an accurate alignment, but if there is enough to throw the caster off, it means the vehicle should visit the frame-straightening shop.

Frame Angle

An element that needs to be considered when performing a wheel alignment is frame angle. If the vehicle is higher in the rear

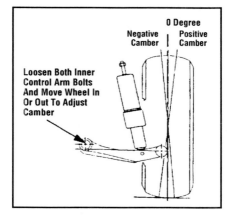

Rear camber adjustment using the inner control arm bolts. *(Courtesy: GM Corp.)*

than in the front, it has a positive frame angle. However, if the vehicle is lower in the rear than in the front, it has a negative frame angle. Most light trucks have a positive frame angle from the factory because they are anticipating that the vehicle will be normally driven with a load. So how does this affect wheel alignment?

As the vehicle is raised up in the rear, the top of the steering axis moves forward, which causes caster to become more negative as a result. Likewise, as a vehicle is lowered in the rear, such as when carrying a heavy load, the top of the steering axis moves rearward, which causes the caster to become more positive.

Taking frame angle into consideration during an alignment and compensating for it will provide the best caster angle when the vehicle is level with a normal load. It is not necessary to compensate for frame angle if the vehicle is normally driven unloaded. However, if the vehicle is driven with a load, it is best to align it under these conditions. If it is not possible to align a vehicle with its normal load due to weight limits or height restrictions, etc., frame angle should be considered when measuring and adjusting caster.

Caster can be compensated for frame angle as follows: Suppose the vehicle has a positive frame angle of 2.0 degrees (as measured at the frame with a protractor) and the vehicle front caster specification is 4.0 degrees. In the case of a positive frame angle, you want to subtract the frame angle from the caster specification to arrive at the desired specification, which in this case is 2.0 degrees. Likewise, if the frame angle is normally negative, you would add it to the caster specification for the vehicle in order to arrive at the desired caster specification. So using the same example as before (2.0 degree frame angle and 4.0 degree caster spec.)

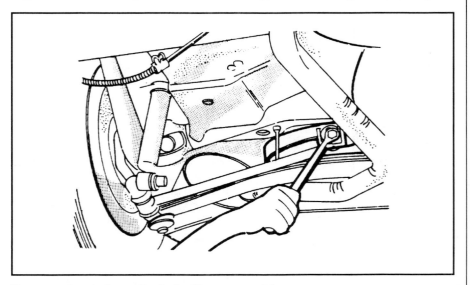

Rear camber being adjusted with an eccentric.

TRAINING FOR CERTIFICATION

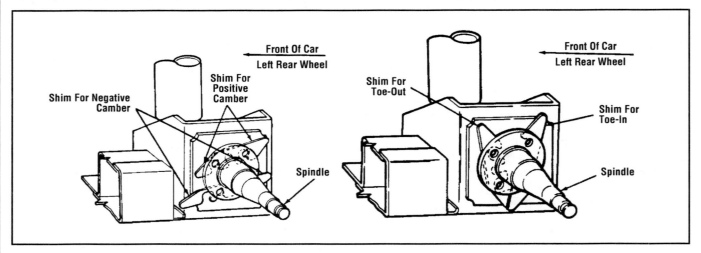

On some vehicles, rear camber and toe are adjusted using shim combinations at the rear spindle. *(Courtesy: DaimlerChrysler Corp.)*

the desired specification is 6.0 degrees. Applying the correct frame angle corrections to an unladed vehicle, that is driven normally loaded, will provide better handling while being driven.

ADJUSTING ALIGNMENT

Before performing an alignment, make sure the vehicle is at its curb weight, which means all fluids are at their capacity, the jack and spare tire are stored in their proper position, and all loads that are not part of the vehicle are removed. An exception to this would be for vehicles like work trucks and vans that are normally driven loaded. Make sure the tires are inflated to their specified cold pressure.

Drive the vehicle onto the alignment rack and set up the alignment equipment according to the manufacturer's instructions. Keep in mind that not all adjustments described here are possible on all vehicles. Consult the vehicle service manual for specific information.

If a four-wheel alignment is being performed, adjust the rear wheel camber and toe first. On some vehicles, rear camber is adjusted by turning an eccentric or loosening the bolts where the lower control arm pivots on the frame, and then moving the control arm. On others, camber is adjusted by loosening the strut-to-knuckle bolts and changing the position of the knuckle in relation to the strut.

Rear toe can be adjusted by using shims at the trailing arm pivot or at the spindle, by turning an adjustment link, which is similar to a tie-rod end, or by loosening the inner control arm bolts and prying in the necessary

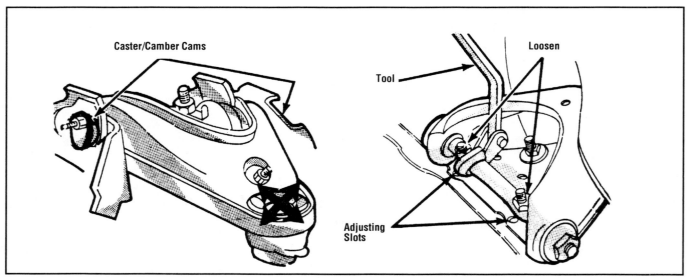

On Short/Long Arm (SLA) type suspensions, caster and camber are adjusted together by moving the pivot points of the upper control arm.

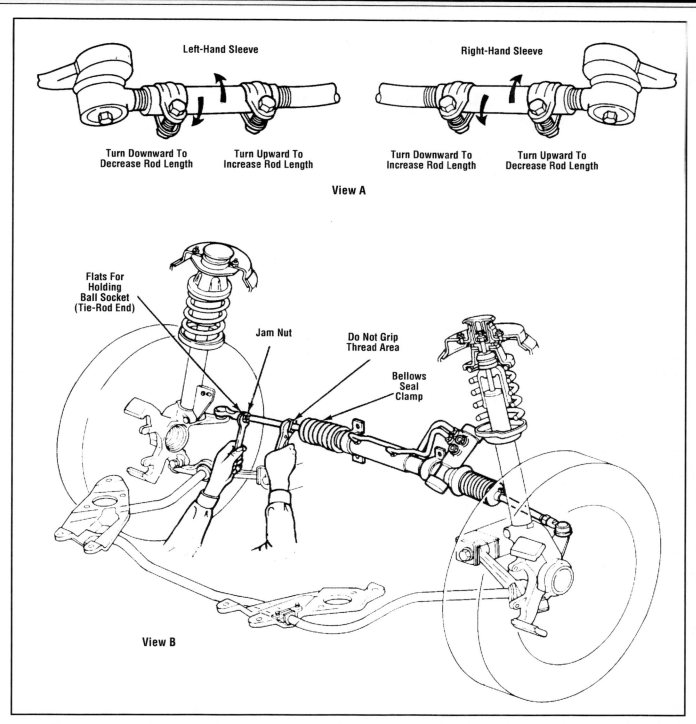

Toe adjustment. View A shows adjustment on conventional steering vehicles while view B shows adjustment on rack-and-pinion steering vehicles.
(Courtesy: Ford Motor Co.)

direction. If the rear toe setting cannot be brought within specification, perform a thrust line alignment.

Measure and correct the front caster and camber. Usually, an adjustment to one of these factors will affect the other, so they are changed together. Caster can be adjusted by moving the pivot points of the upper control arm using eccentrics or shims or by adjusting the strut rod at the lower control arm. Vehicles with MacPherson or modified struts

TRAINING FOR CERTIFICATION

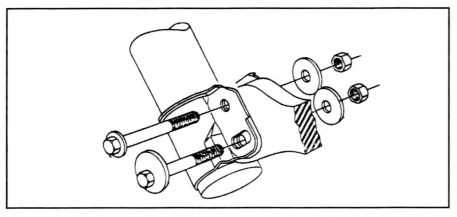

A camber adjustment eccentric at the junction of the strut and steering knuckle.

can be adjusted by moving the upper strut mount in the strut tower.

Camber can be adjusted by moving the pivot points of the upper control arm using eccentrics or shims. Camber on vehicles with MacPherson or modified struts can be adjusted by moving the upper strut mount in the strut tower, or by loosening the strut-to-knuckle bolts and turning an eccentric or moving the knuckle in relation to the strut.

In addition to the above mentioned methods, some trucks use off-center mounted ball joints or eccentric bushings to adjust caster and camber. Sometimes these adjustment devices do not come on the vehicle from the factory, but rather must be installed when adjustment is warranted.

Center and lock the steering wheel and correct the vehicle toe. On vehicles with conventional steering, loosen the tie-rod adjustment sleeve clamp bolts and turn the adjustment sleeve. On vehicles with rack-and-pinion steering, loosen the locknut and turn the inner tie-rod end. Make only half the needed adjustment at each side of the vehicle so the steering wheel will be centered. After toe adjustment, make sure the adjustment sleeve clamps are positioned so they do not contact any suspension members during vehicle operation.

After caster, camber and toe adjustments, check all nonadjustable angles and dimensions as a control for bent or damaged suspension components. Road test the vehicle.

NOTES

TRAINING FOR CERTIFICATION

WHEEL AND TIRE DIAGNOSIS AND REPAIR

TIRE WEAR DIAGNOSIS

If a tire has worn evenly across and around the tread, it has performed properly. Whether the tire has worn out within acceptable mileage depends on the type, weight and load of the vehicle; the inflation pressure; the driving style; and the type and quality of the tire. On many tires, horizontal wear strips will appear when the tread has worn to the legal minimum.

Tire wear patterns are an excellent indicator of suspension problems. Even wear across and around the tread on all the tires indicates the alignment is properly set. Perhaps the most frequent tire wear symptom is symmetrical wear of the sides of the tire while the center tread still has good depth. Ordinarily, this is the result of running for a long time with insufficient air pressure in the tire, allowing the tread patch to lift in the center. Testing the pressure will usually confirm this diagnosis.

Some kinds of suspensions are inclined to produce this wear even without low inflation. Trucks with the Twin I-Beam front suspension will frequently show this pattern on the front tires, regardless of tire pressure, due to 'camber roll'—a greater change of camber throughout suspension travel than with other kinds of front-end geometry. Many vehicles with strut front suspensions show this pattern, too, although this suspension design has very little camber roll. On both types, more frequent tire rotation, front to back and side-to-side, can prolong tire life by somewhat equalizing wear among all four wheels. Wear on the outsides of each tread can also be caused by frequent high-speed cornering.

Excessive wear in the center of the tread, all around the tire, is frequently the result of too-high

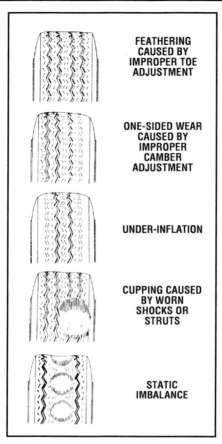

Abnormal tread wear patterns and their causes.

air pressure in the tires. On light trucks and some cars, this can happen if the owner fills the tires to capacity to carry a heavy load, and then forgets to bring the pressure back to normal after removing the heavy load. Excessive pressure also reduces the amount of tread patch touching the ground, and thus the amount of traction and control you can expect through that wheel. It also stiffens the tire so that a larger proportion of small bumps and ruts are transmitted through the suspension.

Sometimes center tread wear is the result of consistently hard acceleration (noticeable on the drive wheels only). Excessive toe-in on independently adjustable rear drive wheels or just lack of rotation of the wheels, especially on the rear wheels of the vehicles mentioned above, could also show outside wear on the front tires.

Other forms of tire wear point to different alignment and suspension problems. If the tire is worn more on one side than on the other, this shows either incorrect camber settings or incorrect toe. Incorrect caster settings may be slightly perceptible as a pull toward the center-that is, difficulty in turning away from straight ahead-or as steering 'wander,' but they generally do not show up as a characteristic wear pattern. On older, bias ply tires, excessive toe-in or toe-out frequently shows as a featheredge on the rubber at either side of the grooves. Radial ply tires show this result less frequently, although from the same cause.

Scalloping or cupping of the tires, in which sections of the tire appear to have been cut out by a sharpened spoon, is the result of wheel imbalance, worn shocks or worn suspension and steering components. Scalloping is more common in larger, heavier tires with thicker treads. Once it begins, the resonance with suspension components will ordinarily cause it to become even worse.

Sharp-edged chunks missing from a tire can result from the impact of road debris on the tire. Such damage may make it impossible to properly balance that wheel. Tires should also be inspected for side bulges (which often indicate broken internal cords), cracks in the sidewalls or treads, and any other damage to the carcass.

Occasionally, tread wear across the tire may be observed. If the patches are even, regular and straight across, these are just the wear indicator marks; although they don't indicate anything amiss in the wear pattern, this means the tire is due for replacement. Somewhat irregular, diagonal wear strips on the rear tires of a front-wheel drive vehicle may be an indication of incorrect rear toe settings. Because of the comparatively lightweight resting on these rear wheels, they ordinarily last for many tens of thousands of miles. But with incorrect rear toe, the tire can hold for part of its revolution and then slip suddenly a bit sideways when rolling tension builds up from the wrong toe setting. Once the wheel finds a place on the tire that it likes to slip, it will do so with each rotation, and the wear pattern will develop.

INSPECTION AND INFLATION

Check the tires for cuts, stone bruises, abrasions, blisters and for objects that may have become embedded in the tread. Most tires have built in tread wear indicators molded into the bottom of the tread grooves. These indicators will appear as 1/2-in. wide bands when the tread depth becomes 1/16-in. When the indicators appear in two or more adjacent grooves, at three locations around the tire, or when any cord or fabric is exposed, the tire must be replaced.

All tires have identification marks pertaining to size, tire pressure and load range molded into the sidewall of the tire. The Department of Transportation grades all tires for tread wear, traction and temperature resistance and this information is also molded into the sidewall. The tread wear grade is expressed as a number and the traction and temperature grades are expressed as the letters A, B or C, A being the best and C being the minimum standard that all tires must meet.

Check tire inflation pressure when the tires are cold, as pressures can increase as much as 6 psi due to heat. The inflation pressure is listed on a decal located on the door jamb.

WHEEL AND TIRE RUNOUT

To check for runout, block the vehicle's wheels and leave the transmission in neutral. Raise the vehicle and support it so you can at least check both wheels on one axle without re-lifting. Check each wheel for wheel bearing play before you measure runout: if the bearings are too loose, there is no way to accurately measure wheel runout on the vehicle. Moreover, on some rear-wheel drive vehicles with tubular rear axles, each axle half can move in and out slightly: be sure your runout test is done with the axle consistently at one end of travel or the other. You want to measure just the wheel's runout, not the axle shaft's range of side-to-side movement (end-play).

There are two wheel runout checks: radial and lateral. However, there are two places to check on the tire and two on the wheel itself. Check the wheel for radial runout inside the rim, just on the outer edge where the tire bead seats. Check it for lateral runout just around the bead corner on the side (don't use the lip of the wheel, as it may not be cut as true as the wheel itself). If available, use the manufacturer's specifications for acceptable wheel runout. As a rule-of-thumb, stamped steel wheels should not be laterally or radially out-of-round more than 0.040-in. (1.0mm), and cast alloy wheels not more than 0.030-in. (0.75mm).

This is also the time to be sure the wheel is mounted correctly, with the hub and lugs exactly centered. On many wheels, it is important to get all the wheel lugs hand-tight before any of them are tightened to their final torque setting.

If this does not correct runout, check the hub and flange. If either of these is bent, no wheel can track straight. Using a dial indicator, check the flange at the machined surface where the lugs attach to the wheel.

Measure the hub at the circular machined surface around the outside of the hub. Flanges and hubs should show almost no measurable runout. Damage to flanges or hubs are the result of forcibly striking the wheel against hard objects, like curbs or parking blocks. If you have to replace a hub, flange or axle, check the wheel bearings carefully, because they will have sustained the same force as the bent component. In many cases, replacement of the hub, flange or axle will require wheel bearing (axle bearing) replacement, too.

Lateral tire runout is measured at a place on the sidewall where raised rubber lettering will not throw off the dial indicator. Radial runout is measured at the center of the tread, though with particularly wide tires you may wish to make runout measurements on both sides. In some cases, internal cord damage will allow a tire to 'grow' runout as wheel speed and centrifugal force increase. This will show up as a wheel that is properly balanced and true at low speeds but falls out-of-balance and out-of-round as the wheel is spun faster. In such a case, the only repair is replacement.

Runout measured at the tire is total runout—the sum of tire runout and wheel runout. The only way to measure tire runout alone would be to mount it on a perfectly round wheel of just the right size and check it. Total

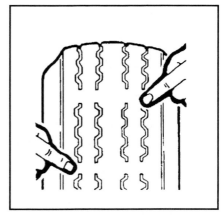

Tread wear indicators will appear when the tire is worn out.

TRAINING FOR CERTIFICATION

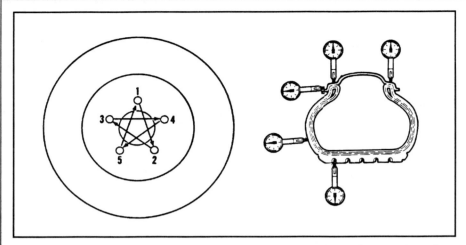

Wheel runout can be isolated to a wheel or tire by a dial indicator placed on each. If the runout is isolated to the wheel, loosen the lug nuts and retighten then evenly in a star pattern to their final torque. If runout is isolated to the tire, reposition the tire on the wheel and rebalance it.

radial and lateral runout, measured at the tire, should not exceed 0.060 in. (1.5mm).

If wheel runout was within specs, but total runout is excessive, the problem is with the tire. If the wheel runout was too much, that will throw off runout measured at the tire, too. Before condemning a tire that is only slightly too far off, try 'match-mounting'—remounting the tire on the rim with the high point of the tire just opposite the high point on the rim. This can often correct the total runout figure. Some tires and wheels have painted or embossed dots indicating match-mount points; check with the manufacturer to find out whether the points should line up or be opposite one another.

If the wheel and tire fail your tests for runout, don't immediately fault them. The first check should be for wheel lug torque. On many lighter vehicles, it is very important that the wheel lugs be tightened in the correct sequence to the specified torque. Over-torquing, under-torquing or even using the wrong tightening sequence can warp the wheel on the hub. Loosen all the lugs, hand-snug them, and re-tighten them in the proper sequence with a torque wrench, then recheck the runout. This simple procedure may bring the wheel back to true.

A tire can pass the runout test when turned slowly on a vehicle on a lift, and still experience runout when driving if there is a large-scale weight imbalance. The effect of the centrifugal force distorts the tire at speed, and runout can occur then (in combination with the vibration caused by the wheel's imbalance).

VIBRATION TROUBLESHOOTING

There are many rotating items in a vehicle's drivetrain, which can cause vibrations. Each part has its own range of natural frequencies. If a natural frequency is excited, the part causes the greatest disturbance as it vibrates. For example, engine vibrations occur at the highest frequency (many hundreds of vibrations per second). This type of vibration causes a buzzing or humming sound and can usually be felt in the floor or dashboard.

However, tire and wheel related vibrations occur at a much lower frequency (10 to 15 Hz). This type of vibration results in the 'shake' most of us have experienced. Even a small amount of wheel imbalance can translate into vibration problems.

Tire-induced vehicle shake is most often caused by the following four items:
• a tire poorly seated on the rim
• an out-of-round tire
• a heavy spot in the tire
• a stiff spot in the tire.

The most critical speed for tire-induced shake is approximately 55-65 mph. At this speed, a tire rotates at 10 to 15 Hz. This is the frequency that corresponds to the natural frequency of a car suspension for virtually all cars, making shake a potential problem for all drivers.

A heavy spot in a tire or tire/rim combination causes a radial force, which bounces the tire up and down (static imbalance) once per tire revolution. This force increases as the vehicle's speed increases. It is only at higher speeds, faster than 40 mph, that this weight-induced bounce becomes noticeable. An out-of-round tire or wheel causes a once per tire revolution up and down force that is independent of speed. If the vibration is not noticeable at low speeds, but begins at greater speeds, tire imbalance is the most likely cause. If a bounce is felt at very low speeds, the most likely cause is an out-of-round tire or wheel.

While the up-and-down shake or bounce results from a heavy spot in the tread, the side-to-side shake or wobble (dynamic imbalance) can result from poor bead seating or a heavy spot in a tire sidewall. If both front tires are affected, the vibration will be added together when both tires are in phase, and may go away completely after a corner, which causes the outside tire to roll more than the inside tire. This tire repositioning can cause the vibrations to cancel each other. However, it is just a matter of time until the tires will be in phase again; then the vibrations will reoccur. This condition can be prevented through proper wheel balancing.

An out-of-round (runout) wheel or tire, or a stiff spot in a tire can cause vibration even in a perfectly balanced tire. In these situations, forces create vibrations that are present regardless of vehicle speed. Wheels don't have force variation, but they all have some runout. Likewise, every tire has some force variation. Each of these variations, by itself, would not cause a problem but there will be a vibration problem if the stiff spot or high point of a tire happens to be placed at the high spot on the rim during mounting. It is easy to avoid this condition by looking for match marks on the wheel and tire. All new tires and most new wheels are marked with dots or similar symbols. When these dots are matched, the high spot on the tire is lined up with the low spot on the wheel, and vibration is minimized. However, if wheels and/or tires are not marked, measure the runout of the wheel and mark the lowest spot on the inside and outside flange. The average low spot of the wheel is midway between these two points. Locate this spot and place a mark on the valve stem side of the wheel. Then mount the tire and align the mark on the new tire with the mark you made on the wheel.

However, if the vehicle is equipped with custom or decorative wheels and you do not want to mar the visible surface of the wheels, you can mount and balance the tire and drive the vehicle. If a vibration is present, dismount the tire and rotate it 180 degrees from its original position on the wheel. This will reduce any vibrations in most cases, but remember to remove the wheel weights and rebalance the wheel before mounting it back on the vehicle. Additionally, keep in mind that worn shocks, struts, or other suspension components can magnify a minor vibration into a large one.

BALANCING WHEELS AND TIRES

Off-Vehicle Balancing

The purpose of balancing wheels is to allow them to rotate at high speeds on the axles without contributing significant vibration of their own to the engine, driveline and road noise sustained by the vehicle. The object is to make the wheel-and-tire combination as equally balanced as possible around its axis of rotation.

Keep in mind that runout and wheel balancing are different operations. A wheel can be balanced, but out-of-round or vice versa. Ordinarily, however, one problem involves the other. Be sure to correct out-of-round before you try to correct balance, or you won't succeed. Both problems will cause similar vibrations on the road, except that out-of-round will be more noticeable at low speeds while out-of-balance will get worse with higher speeds. Often, there is a particular road speed or speeds at which the vibration resonates with the suspension to maximize the imbalance or out-of-round.

Before doing any wheel balancing, you should first check for runout and correct for this, if possible. Next, remove all of the old balancing weights, clean any mud and road tar from the wheel and

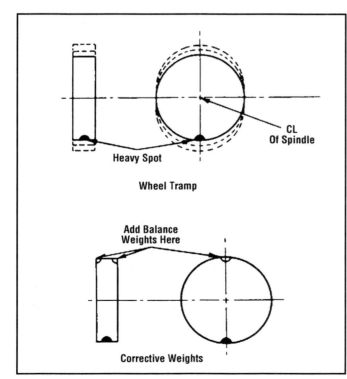

Static unbalance correction. (Courtesy: GM Corp.)

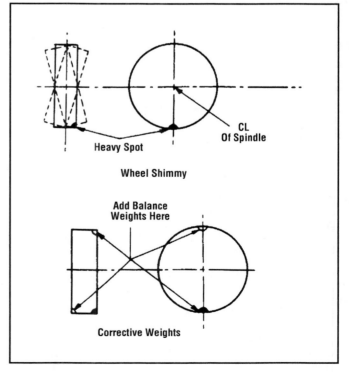

Dynamic unbalance correction. (Courtesy: GM Corp.)

TRAINING FOR CERTIFICATION

remove pieces of gravel and other debris from the tread. Shake the wheel back-and-forth and listen carefully: sometimes there are loose balls of rubber, congealed tire inflation chemicals or even water from infrequently drained compressors rolling around in the cavity. If you hear something inside the wheel or if you're unsure, dismount the tire from the wheel and check. There's no way to balance a wheel with foreign objects bouncing or sloshing around inside, and they will eventually abrade the sealing surface. Be aware that many aerosol tire inflating chemicals are highly flammable, and take proper precautions to keep flames, sparks (including electric motor brushes) and cigarettes away when you deflate a tire.

Wheel balance is divided into two types: static and dynamic. Static balancing is performed as if the tire/wheel was a flat disk balanced at its center point; the heaviest side is identified and measured by a bubble level or other instrument. Once a heavy spot is identified, an equal amount of compensating weights is added to both sides of the rim on the opposite side of the wheel. This makes the weight equal around the disk of the tire/wheel. Static balance is so named because the tire/wheel is kept stationary (static) when the balance is checked, usually on a conical bubble balancer.

In dynamic balancing, we regard the tire/wheel as a somewhat flattened cylinder. This procedure improves on static balancing by recognizing that a wheel can also be off-balance from inside-to-outside, making it wobble more as it rotates at higher speed. Obviously, the wider the wheel and tire, the more important dynamic balancing becomes.

Dynamic balancing is so named because the tire/wheel must be spun to detect any side-to-side imbalance. The wheel-balancing machine then detects any dynamic imbalance and calculates the number and location of

the corrective weights that will be needed. Usually this determination is quite accurate, but there are two potential sources of error.

First, most balancing machines have no way of knowing the internal diameter of a wheel, which is the point where balancing weights will be attached. Instead, the machine is programmed on the assumption of standard sized wheels. Centrifugal force will be different on wheels of different diameters. Therefore, you must measure the wheel and program the wheel-balancing machine prior to spinning the wheel to determine the amount of weights to be added.

The second problem is that the arbors and hubs of wheel balancing machines can also wear, allowing runout and thus imbalance on the machine. You can test for this by balancing a tire/wheel as perfectly as possible, remounting it on the machine in a different position and re-checking its balance. If the balance readout is much different, the variation is in the machine. There will always be some variation, but if it is enough to call for adding, removing or shifting weights, there is too much free-play.

After a wheel is balanced with weights installed, you should spin it up to speed once again to double-check the balance.

There are two kinds of wheel weights: clip-on and adhesive. Clip-on weights are somewhat more precisely made by weight, and the mechanical attachment is generally more secure. You should never use clip-on weights with cast alloy wheels because the clips can scratch the aluminum. Also, there will eventually be electrolytic damage between the steel of the clip and the aluminum of the wheel, particularly in areas where salt is used on the roads in the winter. Not only will the weight fall off, but the wheel may be corroded and damaged as well.

Attach adhesive weights to an aluminum alloy wheel only after you have thoroughly cleaned the

mounting surface of the wheel. Adhesive weights are less precise by weight since they generally come on a strip, and you break off the amount you need. The break is always slightly different, as is the weight.

On-Vehicle Balancing

Some technicians prefer to use on-axle wheel balancing instead of or in addition to other means of balancing the wheels. The advantage of on-axle balancing is that it includes the brake drum or rotor, the hub and the axle when calculating the spinning mass to be balanced. The disadvantages are that it cannot catch inside-outside wheel imbalance, it takes longer, and it requires some special equipment.

Once a wheel has been on-axle balanced, it must be replaced with the same wheel lugs in the same wheel holes, or the balance will be lost. That means, obviously, that the wheels must be spin-balanced on the axle whenever the tires are rotated, a flat tire is replaced with the spare, or a tire is removed from a rim to be patched. On-axle spin balancing is done differently for drive axles and non-drive axles.

To balance non-drive axles, axles that are not turned directly by the powertrain, raise the wheel you are balancing with a floor jack and check for wheel bearing play by shaking the wheel back-and-forth. If the wheel bearing is excessively loose (see manufacturer's technical literature for specs), that problem will have to be corrected before you can balance the wheel on the axle. You can also quickly check for dragging brakes at this point. Besides being a drag-and-pull problem for driving on the road, a dragging brake can overload the balancer's motorized drum while you're spin-balancing a wheel, and burn out its electric motor.

Connect the vibration sensor to the suspension by whatever method the manufacturer recommends. Connect the weight-calculating meter to the sensor if they are separate. Either mark a point on the tire with chalk or use the valve stem for your reference point. Use the motorized drum to spin the wheel up to speed, and release the drum from the wheel. The sensor-meter will flash like an ignition timing light, usually when the heavy side is down, but check your balancer's manual. On most meters, a display will indicate the amount of weight needed. Remember, you must sometimes use slightly more or less weight than the meter indicates because of differences in the diameter of wheels and the effect of centrifugal force.

Opinions vary as to the best way to bring the wheel to a stop. Some motorized drums have an internal brake or a sliding shoe that can stop the wheel. Some shops have the policy of allowing the wheel spin to a stop on its own. While others use the vehicle service brake very gently (otherwise the wheel slams to a stop, which can possibly cause damage to the vehicle or the vibration sensor or perhaps slide the tire or wheel weights on the rim and subsequently falsify the balance measurement). Since on-axle balancing does not distinguish inside-outside imbalance, add half the weight to the inside and half to the outside so your corrective weights are evenly balanced laterally. This won't correct dynamic imbalance, but it won't contribute to it either. Remember to apply the weights opposite the heavy side. With very wide wheels and tires, it is recommended that you first balance them dynamically on the machine, then balance them on-axle only if necessary.

For drive axles, block the non-driven wheels and lift one of the drive wheels. With the transmission in neutral, see if you can easily turn the lifted drive wheel. If there is noticeable friction or if you cannot turn it at all, you have a limited-slip or locking differential. In this case, you will have to raise the opposite wheel also, and remove the wheel and

its brake drum while you balance the first side.

If the opposite wheel has a disc brake and the disc comes off easily, remove that as well. Don't trust a snap-spring washer to hold a disc or drum on. A brake drum or disc that fell off while spinning at 60 mph would become a dangerous missile once it hit the shop floor. What's more, if the drum or disc is unbalanced, that can throw off your balance work on the opposite wheel.

When you balance the wheel on the other side of the axle, remember to mark a mating wheel stud with the appropriate hole in the first side's drum or disc. That way, you can put the brake friction component back in exactly the same position, and keep your balancing work from being lost.

Even if there is no manually-perceptible friction, it is still possible that the differential may be limited-slip or even locking, so check the vehicle owner's manual and manufacturer's information to be sure. A Torsen/Gleason differential, for example, will turn freely from either wheel, but will almost lock when driven by the driveshaft. If there is any reason to suspect you're working with anything but an open differential, raise the opposite wheel to be safe. Otherwise, the vehicle could either damage the differential or drive itself over the wheel chocks and cause an accident.

Remove old wheel balancing weights, mud, lumps of grease or other things that can affect balance. Check the tread for stones or other debris. Listen for foreign objects rolling or sloshing in the tire cavity, and remove them if they are present.

Have an assistant get behind the steering wheel and start the engine. With the balancer sensor and meter set up as for the driven wheel, have your assistant put the transmission in gear and raise the wheel speed. On a vehicle with an open differential and the other tire resting on the ground, he or she should raise the speed to an indicated 30 mph.

Because of the drive-splitting action of the differential, the wheel you are balancing will be turning at 60 mph. This much gear activity is unusual for the differential's spider gears, so perform your balancing work as quickly as is compatible with good workmanship.

If you have a limited-slip or locking differential and have lifted both wheels and removed the wheel and drum opposite, your assistant should raise the speed to an indicated 60 mph, since the drive axles will be turning together at that speed. In this case, too, remove the wheel and disc or drum opposite the side you are balancing while you work, marking the alignment of lugs and holes.

On-vehicle spin balancing is less frequently performed on front-wheel drive vehicles because it is less accurate and more difficult. The chief problem is supporting the raised wheel in such a way that the halfshaft is perfectly straight coming from the transaxle. The operations of some constant velocity (CV) joints may change the balance as the axle rotates and cause balance readings to be inaccurate. The same is true of rear drive axles with independent suspensions and rear axle constant velocity joints. Dynamic balancing is the method of choice for all of these axles.

Be aware that on-axle wheel balancing on a vehicle with a four-wheel anti-lock brake and/or traction control system will probably set a trouble code and interfere with the on-axle balancing procedure. It may even cause the brake system software to disable the anti-lock brake system. Since these systems depend on information from the wheel speed sensors to determine which brake to pulse, spinning one wheel or the drive axle while the other three wheels are stationary may trigger whatever countermeasures that system uses. Learn how to disable or reset these systems from a service manual before you

TRAINING FOR CERTIFICATION

do an on-axle-balancing job on these vehicles.

TIRE ROTATION

The most effective way to get optimal mileage from a set of tires on a properly aligned vehicle is to maintain proper tire pressure; to drive without rapid acceleration, braking or cornering; and to rotate the tires regularly. Automobile manufacturers call for rotating tires every 6,000 to 8,000 miles. Some tire manufacturers recommend intervals as short as every 5,000 miles.

Rotation minimizes tire wear by shifting the tires to new suspension locations. This means the vehicle weight and unique wear patterns generated by the suspension in that corner of the vehicle are canceled out by those in the other corners over the course of several sequential tire rotations. The total amount of wear is the same, but since it is equally spread over all the tires, maximum life is achieved. The purpose of rotation is to prolong the wear and traction of the entire set of tires, so many professionals recommend at least enough tire rotation intervals to put each tire at each corner of the vehicle an equal amount of time during its useful life. However, as we will see, there are some exceptions.

Different vehicles and different tires require different rotation patterns. Check the vehicle manufacturer's technical literature or the owner's manual to be certain.

Certain general guidelines can be used in the absence of other information. For most vehicles, move the back tires to the front and the front to the rear.

Some tires are directional, to channel water out from under the tread. To keep from defeating the purpose of the water-clearing directional tread, these tires can only be switched from front to rear, not left to right (unless they are removed and reinstalled reversed on the wheel), because that would change the direction of tread rotation on the road. In addition, there are a few radial tires that the tire makers say should always be mounted on the same side of the vehicle, regardless of tread. Again, check the manufacturer's literature.

Finally, some high-performance vehicles use different sized tires and wheels for front and rear, ordinarily employing the wider tires on the drive wheels in back. Front-to-back tire rotation is impossible on such vehicles because of the difference in rim and tire width, and since many of these tires are directional, they can't be switched from side-to-side, either. On such vehicles, the only countermeasure to tire wear is the unwelcome advice of conservative driving, and the only correction is replacement.

On light and medium trucks, the spare is often included in the rotation sequence. Check with the vehicle owner's manual to see what pattern is recommended.

TIRE MOUNTING

The connection between the tire and the wheel is, mechanically, an interference fit-that is, the tire opening is slightly smaller than the wheel that fits in it. The high friction resulting from this fit, along with the pressure, holds the tire in place and seals the air inside the tire. The tire mounting machine, of course, stretch-wraps the tire around the wheel, and air pressure forces the beads onto the portion of the rim that is tight enough to seal the air inside.

For safety, when seating tires on rims, never exceed the maximum pressure specification set by the tire manufacturer for the tire. If this pressure is insufficient to seat the bead, break the bad seal, clean the rim and the tire and slightly lubricate the mating surfaces. Thirty psi should be enough to seat almost all tires unless there is another problem you should solve first.

Often it is necessary to treat tires with a special tire lubricant to get them to fit over a wheel or to seat in the rim. Use of the lubricant should be kept to a minimum because it may contain moisture that could eventually cause rust inside the wheel.

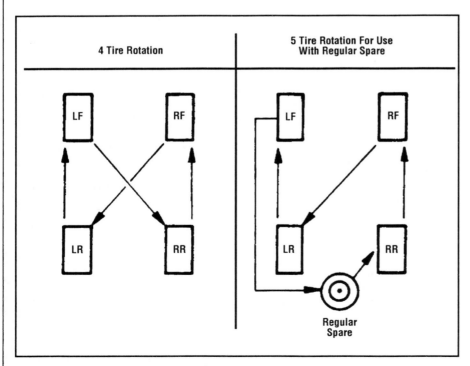

Typical tire rotation patterns. *(Courtesy: Ford Motor Co.)*

NOTES

NOTES

NOTES

Prepare yourself for ASE testing with these questions on
SUSPENSION AND STEERING

	Readings		Specs
	Left	Right	Left or Right
Camber	+3/4° or +45 min.	-1 1/2° or -1°30 min.	0 to +1/2° or 0 to +30 min.
Caster	0°	0°	0° to +1°
Toe-In	1/16 in. or .16mm		1/16 in. to 3/16 in. or .16mm to .48mm

1. The above alignment settings would result in which of the following conditions?
   A. left tire wear on the inside, vehicle does not pull to either side
   B. right tire wear on the inside, vehicle pulls to the left
   C. right tire wear on the outside, vehicle pulls to the left
   D. right tire wear on the outside, left tire wear on the inside and vehicle pulls to the left

2. A driver says that the front end of his car vibrates up-and-down while traveling at most road speeds. Technician A says that too much runout of the front wheels could be the cause. Technician B says that static out-of-balance of the front wheels could be the cause. Who is right?
   A. Technician A only
   B. Technician B only
   C. Both A and B
   D. Neither A or B

3. A vehicle has front wheel shimmy at low speed and requires increased steering effort. Technician A says that too much positive camber could be the cause. Technician B says that too much positive caster could be the cause. Who is right?
   A. Technician A only
   B. Technician B only
   C. Both A and B
   D. Neither A or B

4. All of these could cause tire wear if not within manufacturer's specs **EXCEPT**:
   A. caster
   B. wheel balance
   C. toe-in
   D. camber

5. A vehicle with electronically controlled power steering has a lack of power assist in both directions. Technician A says that the problem is in the electronic control system. Technician B says that there could be a problem in the steering gear or a bad power steering pump. Who is right?
   A. Technician A only
   B. Technician B only
   C. Both A and B
   D. Neither A or B

6. A customer complains that a vehicle wanders on a bumpy road and needs continual driver correction to stay straight. What could be wrong?
   A. excessive negative front caster
   B. insufficient front toe
   C. rear or front bump steer
   D. all of the above

7. While aligning a front end, a technician finds the toe-out on turns (turning radius) to be greater than manufacturer's spec and it has 2.0 degrees difference from the left side to the right side. Which, if any, of the following could be the cause?
   A. nothing is wrong; the vehicle has a non-symmetrical steering design
   B. bent pitman arm
   C. bent tie-rod
   D. both B and C

8. When checking a tire, a technician finds it to have too much axial (up-and-down) runout. Technician A says to rotate the tire and wheel assembly on the mounting studs to correct the problem. Technician B says to rotate the tire on the wheel assembly to attempt to bring the runout within specifications. Who is right?
   A. Technician A only
   B. Technician B only
   C. Both A and B
   D. Neither A or B

9. If you readjust camber on the left front strut, which, if any, of these specifications might change?
   A. right caster
   B. left toe and total toe
   C. right SAI
   D. none of the above

10. Too much caster on the left front wheel will cause the _____?
- A. vehicle to drift to the left
- B. left tire to wear on the outside edge
- C. vehicle to drift to the right
- D. left tire to wear on the inside edge

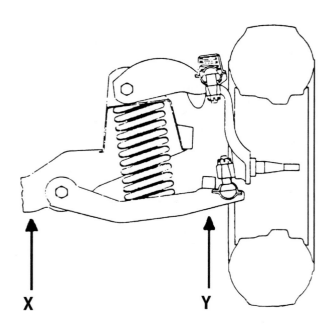

11. The suspension shown above is being checked for ball joint wear. Technician A says that the inspection can be made with the front end supported at point X. Technician B says that the inspection can be performed with the front end supported at point Y. Who is right?
- A. Technician A only
- B. Technician B only
- C. Both A and B
- D. Neither A or B

12. A pressure test is being performed on a car with power steering. The pressure readings taken when the wheels are at the right and left stops are below specs. The tester readings are normal when the tester shutoff valve is closed. Technician A says that these readings could be caused by a bad steering gear. Technician B says that a bad pump could cause these readings. Who is right?
- A. Technician A only
- B. Technician B only
- C. Both A and B
- D. Neither A or B

13. A vehicle with a manual rack-and-pinion steering-gear has a shimmy. Technician A says that worn rack-to-frame mounting bushings could be the cause. Technician B says that loose inner or outer tie-rod ends (sockets) could be the cause. Who is right?
- A. Technician A only
- B. Technician B only
- C. Both A and B
- D. Neither A or B

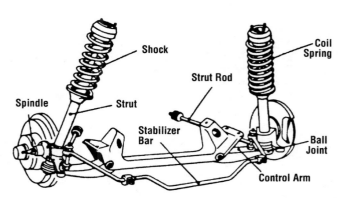

14. The ball joints on the MacPherson strut-type suspension shown above are being replaced. Technician A says that a coil spring compressor should be installed before separating the control arm and the spindle. Technician B says that the wheel alignment should be checked after the ball joints have been replaced. Who is right?
- A. Technician A only
- B. Technician B only
- C. Both A and B
- D. Neither A or B

15. The rear thrust line reads to the left (driver's) side while the vehicle is at normal ride height. If the wheels were pointing straight ahead, which direction would the vehicle tend to wander?
- A. veer to the left
- B. veer to the right
- C. wander to both left and right
- D. no change occurs

16. Technician A says that the bearing preload must be set when overhauling a manual steering gear. Technician B says that the sector lash must be set when overhauling a manual steering gear. Who is right?
- A. Technician A only
- B. Technician B only
- C. Both A and B
- D. Neither A or B

Prepare yourself for ASE testing with these questions on SUSPENSION AND STEERING

17. A vehicle with power steering has assist in only one direction. Technician A says that oil leaking past a worm shaft seal ring could be the cause. Technician B says a broken ring on the worm piston could be the cause. Who is right?
   - A. Technician A only
   - B. Technician B only
   - C. Both A and B
   - D. Neither A or B

18. A vehicle with power steering has no assist in either direction. Technician A says that a loose power steering belt could be the cause. Technician B says that a faulty flow control valve could be the cause. Who is right?
   - A. Technician A only
   - B. Technician B only
   - C. Both A and B
   - D. Neither A or B

19. A customer complains of excessive body roll. Which of the following components is used to control body roll?
   - A. stabilizer bar
   - B. ball joints
   - C. coil springs
   - D. body mounts

20. Technician A says that the ride height of a vehicle is always adjustable. Technician B says that the ride height of a vehicle can only be adjusted on certain models. Who is right?
   - A. Technician A only
   - B. Technician B only
   - C. Both A and B
   - D. Neither A or B

21. Technician A says that caster is not a tire-wearing angle. Technician B says that toe-in or toe-out will cause tire wear. Who is right?
   - A. Technician A only
   - B. Technician B only
   - C. Both A and B
   - D. Neither A or B

22. Technician A says that if steering axis inclination is out-of-specification, a bent control-arm may be the cause. Technician B says that if steering axis inclination is out-of-specification, the spindle can be shimmed to correct the problem. Who is right?
   - A. Technician A only
   - B. Technician B only
   - C. Both A and B
   - D. Neither A or B

23. Tire cupping is most likely caused by _____?
   - A. worn shock absorbers
   - B. wheel imbalance
   - C. loose tie-rod ends
   - D. all of the above

24. Which of the following is **MOST** likely to be lubricated and sealed by the manufacturer?
   - A. needle bearings
   - B. tapered roller bearings
   - C. plain bearings
   - D. ball bearings

25. Excessive steering wheel play is felt in the steering wheel on a manual steering system. Technician A says that a worn idler arm could be the cause. Technician B says that a maladjusted mesh preload may be the cause. Who is right?
   - A. Technician A only
   - B. Technician B only
   - C. Both A and B
   - D. Neither A or B

26. A vehicle suspension system is being checked for radial and axial tolerances. Technician A says the ball joints should be checked. Technician B says that the idler arm should be checked. Who is right?
   - A. Technician A only
   - B. Technician B only
   - C. Both A and B
   - D. Neither A or B

27. A vehicle has a noise coming from the power steering pump. Technician A says that a loose belt could be the cause. Technician B says that air in the system could be the cause. Who is right?
    A. Technician A only
    B. Technician B only
    C. Both A and B
    D. Neither A or B

28. All of these are types of power steering pumps, **EXCEPT**:
    A. vane
    B. roller
    C. slipper
    D. crescent

29. All alignment measurements on a vehicle are within specification, except for SAI and toe-out on turns. Technician A says that these alignment angles are not adjustable, so there is nothing more to do. Technician B says neither of these angles can cause tire wear. Who is right?
    A. Technician A only
    B. Technician B only
    C. Both A and B
    D. Neither A or B

30. A car is raised on a lift with the wheels pointed straight ahead and when the engine is started, the wheels slowly turn to the right. Technician A says that a binding upper bearing on a MacPherson strut or an improperly installed RBS tie-rod end could cause this condition. Technician B says the power steering valve is out of adjustment. Who is right?
    A. Technician A only
    B. Technician B only
    C. Both A and B
    D. Neither A or B

31. On a 4-wheel drive vehicle with drag link steering, after the alignment is complete and correct, the steering wheel is turned to the left when the car is driving straight ahead. Technician A suggests removing the steering wheel and replacing it in an orientation closer to straight-ahead. Technician B suggests advising the driver that the linkage does not include enough adjustment to center the steering wheel. Who is right?
    A. Technician A only
    B. Technician B only
    C. Both A and B
    D. Neither A or B

32. A tire on an alloy wheel is being balanced. Technician A says that clip-on weights should be used. Technician B says that tape on weights should be used. Who is right?
    A. Technician A only
    B. Technician B only
    C. Both A and B
    D. Neither A or B

33. A customer is purchasing new front tires for her front-wheel drive vehicle. Technician A says the best way to balance the tires is to mount them on the vehicle and use an on-axle spin balancer. Technician B says the best way to balance the tires is to balance them on a spin balancing machine prior to installation. Who is right?
    A. Technician A only
    B. Technician B only
    C. Both A and B
    D. Neither A or B

34. When the suspension is worked up-and-down on a vehicle equipped with MacPherson strut-type suspension, the camber specification changes more on one side than the other. Technician A says that camber can't be adjusted on a MacPherson strut. Technician B says that the strut on the side that camber changes the most may be bent. Who is right?
    A. Technician A only
    B. Technician B only
    C. Both A and B
    D. Neither A or B

35. A vehicle equipped with power steering exhibits excessive steering effort. All hydraulic system specifications are correct and the alignment is within specifications. Technician A says that this condition may be caused by corrosion in the steering linkage ball studs. Technician B says that the intermediate shaft U-joints could be binding. Who is right?
    A. Technician A only
    B. Technician B only
    C. Both A and B
    D. Neither A or B

Prepare yourself for ASE testing with these questions on
SUSPENSION AND STEERING

36. A medium-duty light truck is on the alignment rack. Technician A points out that there is no arch at all in the rear leaf springs and says this indicates that the springs are either worn out or broken. Technician B says that this is normal on newer trucks and is not a cause for concern. Who is right?
- A. Technician A only
- B. Technician B only
- C. Both A and B
- D. Neither A or B

37. A vehicle equipped with a MacPherson strut-type suspension is in the shop for new tires and an alignment. The tire installer noticed that the vehicle has some play in the left-side inner tie-rod end and informs the alignment technician. The alignment technician should tell the customer _____?
- A. It is not a problem that will affect alignment.
- B. He'll align the front end, but they should replace the inner tie-rod end soon.
- C. The inner tie-rod end needs to be replaced before an alignment can be performed.
- D. He can set the vehicle's alignment so that the play in the inner tie-rod will be compensated for.

38. A 1-ton pick-up truck is in for an alignment. The pick-up truck usually carries some pipes and ladders on a rack attachment on the pick-up's bed. These items need to be removed in order to raise the truck on the alignment rack. After removing the items, the alignment technician realizes that the frame angle has increased by 1.5 degrees. The caster specification for the vehicle is 3.0 degrees. Technician A says that the caster should be adjusted to 1.5 degrees in order to provide the best handling when the vehicle is loaded. Technician B says that the caster should be adjusted to 4.5 degrees in order to provide the best handling when the vehicle is loaded. Who is right?
- A. Technician A only
- B. Technician B only
- C. Both A and B
- D. Neither A or B

39. A vehicle just had new sway bar links and bushings installed. Technician A says that the wheels need to be straight-ahead with the front end supported before tightening the link's fasteners. Technician B says the vehicle should be resting level on its wheels at its normal ride height before the link's fasteners are tightened. Who is right?
- A. Technician A
- B. Technician B
- C. Both A and B
- D. Neither A or B

40. A vehicle equipped with manual rack-and-pinion steering is having the front end inspected. Technician A says that the inner tie-rod ends should be inspected while in their normal running position. Technician B says that if movement is felt between the tie-rod stud and the socket while the tire is moved in-and-out, the outer tie-rod end should be replaced. Who is right?
- A. Technician A only
- B. Technician B only
- C. Both A and B
- D. Neither A or B

41. Which of the following four-wheel alignment inspection and measurement sequences is correct?
- A. ride height, rear camber, rear toe, front caster
- B. ride height, front caster, front camber, front toe
- C. rear toe, rear camber, front caster, front camber
- D. ride height, rear toe, rear camber, front caster

42. A vehicle with power steering is parked with the engine running. A hissing noise is heard when the steering wheel is turned back-and-forth. Technician A says that the power steering pump could be bad. Technician B says that this sound is part of normal operation. Who is right?
- A. Technician A only
- B. Technician B only
- C. Both A and B
- D. Neither A or B

43. All of the following could be caused by a misaligned subframe **EXCEPT**:
- A. difficulty shifting the transmission
- B. hard steering
- C. tire wear
- D. poor ride quality

44. All of the following suspension components are always replaced as a unit **EXCEPT**:
- A. tie-rod end
- B. idler arm
- C. pitman arm
- D. ball joint

86

45. At the beginning of a power steering system pressure test, with the shutoff valve open, the pressure reading is above specification with the fluid at normal operating temperature and the front wheels in the straight-ahead position. Technician A says that there could be a restriction in the power steering pressure hose. Technician B says that the poppet valve in the steering gear could be the cause. Who is right?
  A. Technician A only
  B. Technician B only
  C. Both A and B
  D. Neither A or B

46. Technician A says that dynamic wheel balancing can be accomplished with a bubble balancer. Technician B says that static wheel balancing requires a spin balancer. Who is right?
  A. Technician A only
  B. Technician B only
  C. Both A and B
  D. Neither A or B

47. All of the following will cause a tire to wear at the edge **EXCEPT**:
  A. incorrect camber
  B. overinflation
  C. aggressive cornering
  D. underinflation

48. All of the following can cause incorrect vehicle ride height **EXCEPT**:
  A. a broken leaf spring shackle
  B. a fatigued coil spring
  C. incorrectly adjusted torsion bars
  D. worn shock absorbers

49. Technician A says that lubricant can be used to help seat a tire on a wheel rim. Technician B says that if necessary, it is OK to exceed the maximum tire pressure rating to seat the bead, as long as pressure is reduced immediately afterward. Who is right?
  A. Technician A only
  B. Technician B only
  C. Both A and B
  D. Neither A or B

50. Shock absorbers are being installed in the rear of a rear-wheel drive solid axle vehicle with coil springs. Technician A says the axle must be supported before the shocks are removed. Technician B says that the air must be purged from the shocks prior to installation. Who is right?
  A. Technician A only
  B. Technician B only
  C. Both A and B
  D. Neither A or B

51. A technician has just installed a replacement for a failed power steering gear. Which of the following should he do next?
  A. bleed the system
  B. pressure test the system
  C. flush the system
  D. check wheel alignment

52. Technician A says that setback is caused by collision damage. Technician B says that setback can be designed into a vehicle. Who is right?
  A. Technician A only
  B. Technician B only
  C. Both A and B
  D. Neither A or B

53. Technician A says that a power steering pump pulley should be pressed onto the pump shaft until it stops, to ensure proper pulley alignment. Technician B says that pulley replacement requires that the power steering pump be removed from the vehicle. Who is right?
  A. Technician A only
  B. Technician B only
  C. Both A and B
  D. Neither A or B

54. All of the following could warrant replacement of a rack-and-pinion inner tie-rod end **EXCEPT**:
  A. torn bellows boot
  B. failed articulation test
  C. socket looseness
  D. incorrect rack bearing preload

Prepare yourself for ASE testing with these questions on
SUSPENSION AND STEERING

55. Technician A says that all vehicles with strut-type front suspension require that the strut be removed to replace the front coil springs. Technician B says that the coil spring should be compressed before removal from an SLA-type front suspension. Who is right?
- A. Technician A only
- B. Technician B only
- C. Both A and B
- D. Neither A or B

56. All of the following are methods used by manufacturers to attach ball joints to control arms **EXCEPT**:
- A. riveted on
- B. welded on
- C. threaded on
- D. pressed on

57. All of the following are true of torsion bars **EXCEPT**:
- A. They can be mounted longitudinally or transversely.
- B. They serve the same function as coil springs.
- C. They are interchangeable from side-to-side.
- D. They can be used to adjust ride height.

58. Which of the following causes for ball joint replacement would also require steering knuckle replacement?
- A. torn dust cover
- B. worn ball and socket
- C. wear indicator below surface
- D. broken ballstud

59. Technician A says that, when on-vehicle balancing a drive axle tire, the opposite wheel must also be supported off the ground if the vehicle is equipped with a limited-slip differential. Technician B says that rotating on-vehicle balanced tires presents no special problems. Who is right?
- A. Technician A only
- B. Technician B only
- C. Both A and B
- D. Neither A or B

60. Technician A says that vehicle ride height is measured between the ground and the rocker panel. Technician B says that ride height is measured between suspension components. Who is right?
- A. Technician A only
- B. Technician B only
- C. Both A and B
- D. Neither A or B

61. All of the following are sensors in an electronically controlled suspension system **EXCEPT**:
- A. wheel speed sensor
- B. wheel position sensor
- C. vehicle speed sensor
- D. height sensor

62. A customer complains that his vehicle stalls repeatedly while parallel parking. Technician A says that the IAC valve is the most likely cause. Technician B says that the power steering pressure sensor is at fault. Who is right?
- A. Technician A only
- B. Technician B only
- C. Both A and B
- D. Neither A or B

63. An upper control arm is being removed from a late model vehicle with SLA suspension. Technician A says that the lower arm should be supported with a jack stand close to the ball joint. Technician B says that the coil spring must be compressed with a spring compressor. Who is right?
- A. Technician A only
- B. Technician B only
- C. Both A and B
- D. Neither A or B

64. A vehicle pulls to the right during braking. Technician A says that the problem is in the brake system, and the pull is most likely due to brake fluid on the linings or a seized caliper piston. Technician B says that the pull could be caused by worn right side strut rod bushings. Who is right?
- A. Technician A only
- B. Technician B only
- C. Both A and B
- D. Neither A or B

65. On a vehicle with conventional steering gear and parallelogram steering linkage, an incorrect toe setting can be caused by all of the following **EXCEPT**:

    A. worn idler arm bushings

    B. incorrect tie-rod adjustment

    C. bent center link

    D. too little worm bearing preload

NOTES

NOTES

Answers to Suspension & Steering Test Questions

1. The correct answer is B. The vehicle will always pull towards the side with the most positive camber setting. Since the right side has too much negative camber, this will cause the tire to tilt inwards at the top and cause the right tire to wear the inside shoulder of the tire more quickly. It is also likely that the left tire would wear on the outside shoulder due to too much positive camber.

2. The correct answer is C. Too much runout or static out-of-balance in the front wheels will cause a vibration.

3. The correct answer is B. Excessive positive caster will increase steering effort, cause the steering wheel to return too rapidly, cause wander at high speed and shimmy at low speed.

4. The correct answer is A. Caster that is out of specification will affect steering and handling characteristics, but will not cause tire wear. Whereas, too much toe-in or out would severely wear the tires. Additionally, excessive camber would cause inside or outside shoulder wear on the tire, depending on whether the camber was too negative or positive.

5. The correct answer is B. Although further diagnosis will be necessary, technician B is right. Technician A is wrong because, when there is a malfunction in an electronically controlled steering system, the system will allow full power steering.

6. The correct answer is D. Excessive negative caster, insufficient front toe, front or rear bump steer could cause a vehicle to wander on a bumpy road.

7. The correct answer is D. Either a bent pitman arm or a tie-rod could cause the specifications for toe-out on turns to be out of specification. Vehicles with non-symmetrical steering systems will have a difference from side-to-side, but the difference will not exceed manufacturer's specifications and it rarely has greater than 1.5 degrees difference from side-to-side.

8. The correct answer is B. Rotating the tire on the wheel will compensate for wheel/tire differential (e.g., positioning the highest spot on the tire at the lowest spot on the wheel). Whereas, rotating the wheel assembly on the mounting studs will have no effect on runout.

9. The correct answer is B. Readjusting camber will have an affect on the individual toe for the left wheel as well as the total toe. SAI is not adjustable and generally, caster is not adjustable either on a strut suspension.

10. The correct answer is C. A vehicle will tend to steer or drift to the side with the least amount of caster. Caster will not cause tire wear, but excessive camber will cause wear to the inside or outside shoulders of a tire.

11. The correct answer is B. In order to unload the ball joints, the vehicle should be raised at point Y so that the spring is compressed. Raising the vehicle at point X will still allow the ball joint to be loaded and the inspection results will be inaccurate.

12. The correct answer is A. A bad steering gear would cause the readings to be low when the steering is at the right and left stops, but the readings would be normal with the tester shutoff valve closed.

13. The correct answer is C. Steering play or looseness may occur if the rack-to-frame mounting bushings are worn, or if the inner or outer tie-rod ends are loose. Both of these conditions can result in a shimmy.

14. The correct answer is B. The wheel alignment should be checked after the ball joint (or any suspension part) is replaced. It is not necessary to install a coil spring compressor prior to separating the control arm and the spindle because the MacPherson strut assembly contains the spring. However, a spring compressor should be used if the strut is being disassembled.

15. The correct answer is B. If the thrust line is to the left and the steering wheel is held straight, the vehicle will veer to the right. This happens because the rear wheels will travel towards the left, which will force the front of the vehicle to the right.

16. The correct answer is C. Both bearing preload and sector lash adjustments are necessary after an overhaul. Correct sector lash prevents gear binding and excessive steering wheel free-play. Correct preload prevents worm end-play and loose steering feel.

17. The correct answer is C. An oil leak past the worm shaft oil sealing ring or a broken ring on the worm piston could cause the power steering to only have assist in one direction.

18. The correct answer is C. A faulty control valve will reduce the fluid pressure and cause an increase in steering effort. A loose power steering pump drive belt will create a similar condition.

19. The correct answer is A. The stabilizer bar, also known as the sway bar, is used to control body roll. The ball joints, body mounts and coil springs affect ride and handling but have no effect on body roll.

20. The correct answer is B. Ride height is only adjustable on some models, such as those equipped with torsion bars. On most models, parts replacement is the only way to correct ride height.

21. The correct answer is C. Caster will have an effect on steering and handling, but will not cause tire wear. However, if a suspension system has too much toe-in or toe-out, tire wear will result.

22. The correct answer is A. A bent control arm would cause the angle of pivot points to change and subsequently change the steering axis inclination angle. Technician B is incorrect because shimming a spindle is not a recommended procedure and it would affect the included angle, not the steering axis inclination angle.

23. The correct answer is D. Tire cupping or scalloping can result from worn shock absorbers, improperly balanced wheels/tires or loose/defective tie-rod ends.

24. The correct answer is D. Most ball bearings are factory-lubricated and sealed for life.

25. The correct answer is C. Excessive steering wheel play may result from a worn idler arm and/or a maladjusted mesh preload.

26. The correct answer is A. When checking the radial and axial tolerances, an inspection of the ball joints should be performed. The idler arm would not affect radial or axial tolerances.

27. The correct answer is C. A loose power steering pump belt and air could both cause power steering pump noise.

28. The correct answer is D. A power steering pump can be a roller, slipper or vane type.

29. The correct answer is D. While it is true that SAI and toe-out on turns cannot be adjusted, they indicate that some component is bent and must be replaced. Incorrect SAI will cause changes in camber while turning and incorrect toe-out will abrade tire rubber.

30. The correct answer is B. A binding upper bearing or incorrectly installed RBS tie-rod end would pull the steering to a specific point and stop regardless of whether the engine was running or not. Since the engine needs to be running to have this condition occur, the power steering valve is causing the problem.

31. The correct answer is C. Many steering systems with this type linkage geometry do not have fully adjustable toe links, but the vehicles have steering wheel and column splines that allow multiple orientations of the steering wheel to the column.

32. The correct answer is B. Never use clip-on weights with cast alloy wheels because the clips can scratch the aluminum. Also, there will eventually be electrolytic damage between the steel of the clip and the aluminum of the wheel, particularly in areas where salt is used on the roads in the winter. Not only will the weight fall off, but the wheel may be corroded and damaged as well.

33. The correct answer is B. On-axle spin balancing is usually not performed on front-wheel drive vehicles because it is less accurate and more difficult. The chief problem is supporting the raised wheel in such a way that the halfshaft is perfectly straight coming from the transaxle. The operations of some constant velocity (CV) joints may change the balance as the axle rotates and cause balance readings to be inaccurate.

34. The correct answer is B. While some MacPherson strut assemblies do not allow camber or caster adjustments, many vehicles do have eccentrics or slotted mounting holes to provide adjustment. An abnormal change in camber normally indicates a bent strut.

Answers to Suspension & Steering Test Questions

35. The correct answer is C. Worn dust covers on ball studs can allow contamination and rusting of ball studs in their sockets. However, some vehicles have an intermediate shaft connecting the steering column with the gear. The U-joints on this shaft are not serviceable and are subject to rust and binding.

36. The correct answer is B. Advancements in metallurgy have allowed leaf springs to be manufactured with almost no arching, which allows for lower profiles and lighter weight without sacrificing load capacity.

37. The correct answer is C. Play in the inner tie-rod end will not allow a correct alignment to be performed. The tie-rod end must be replaced prior to performing an alignment.

38. The correct answer is A. If a vehicle is driven normally loaded, the frame angle should be considered when measuring and adjusting caster. In the case of a positive frame angle, the amount of frame angle should be subtracted from the caster specification. If the vehicle is not normally loaded while being driven, it is not necessary to consider frame angle when measuring and adjusting caster.

39. The correct answer is B. A vehicle should be resting on its wheels at its normal ride height when the sway bar link bushing fasteners are tightened in order to get the longest life out of the new bushings.

40. The correct answer is C. Inner tie-rods should be inspected while in their normal running position. This allows the tie-rod assemblies to be level and prevents binding that may occur when the suspension is allowed to hang free. Any play felt between the tie-rod stud and the socket while the tire is moved in-and-out, indicates that the assembly is worn and requires replacement.

41. The correct answer is A. Ride height must be measured and corrected before an alignment is performed. When performing a four-wheel alignment, the rear must be adjusted first beginning with the camber adjustment.

42. The correct answer is B. A hissing noise, most often heard when the wheels are turned and the vehicle is not moving, is caused by normal relief valve operation and is not an indicator of a power steering system problem.

43. The correct answer is D. Many vehicles use subframes to mount the drivetrain and/or suspension and steering. If the subframe is misaligned due to incorrect installation after service or due to collision damage, it can cause binding in the shift and steering linkage. The alignment could also be incorrect which could cause tire wear. Ride quality would most likely be unaffected.

44. The correct answer is B. Although most idler arms are replaced as a unit, on some the bushings can be serviced. Ball joints, tie-rod ends and pitman arms are always replaced.

45. The correct answer is C. Either one could be the cause for the above normal pressure.

46. The correct answer is D. A bubble balancer is used to static balance a tire, while a spin balancer is used to dynamic balance a tire.

47. The correct answer is B. Overinflation will cause excessive wear in the center of the tread.

48. The correct answer is D. Misadjusted torsion bars, sacked coil springs and broken leaf spring shackles would all cause a vehicle's ride height to be too low. Worn shock absorbers would not affect vehicle ride height.

49. The correct answer is A. When seating tires on rims, never exceed the maximum pressure specification set by the tire manufacturer for the tire.

50. The correct answer is C, both technicians are correct. In this suspension design, the shocks are under tension from the coil spring when the vehicle is raised and the suspension is at full rebound, therefore, the axle must be supported before the shocks are removed. Before installing the new shocks, they must be purged of air. With the shock absorber right side up, extend it fully, then turn it upside down and fully compress it. Repeat the procedure three more times to make sure any trapped air has been expelled.

51. The correct answer is C. When the steering gear failed, it most likely contaminated the whole system. The system should be flushed, and then air bled. Toe may need adjusting due to the new gearbox, but a pressure test will only be necessary if there are any more problems.

52. The correct answer is C, both technicians are correct. Ordinarily, if you find setback on a vehicle, it means there has been frame or unibody damage that has shifted the mounting points of the suspension components. However, on a few vehicles, such as light trucks with Twin I-Beam front suspension, the front wheel is slightly farther back on one side by design.

53. The correct answer is D, neither technician is correct. Pulley alignment should be checked as the pulley is installed; many pulleys are not installed until they bottom on a shoulder but rather just until they align. Many power steering pumps cannot be removed from the vehicle until the pulley is removed.

54. The correct answer is D. A torn bellows boot would let dirt and grit get to the tie-rod end socket and could cause it fail. Looseness in the socket or failing an articulation effort test would require replacement. Incorrect rack bearing preload would not affect the inner tie-rod end.

55. The correct answer is B. The coil spring on a Short/Long Arm (SLA) suspension vehicle should be compressed with a suitable spring compressor before the ball joint is disconnected, so it can safely be removed. Technician A is wrong because on vehicles with modified struts, the coil spring is located between the lower control arm and frame, as on an SLA suspension vehicle, and the strut need not be removed to replace the coil spring.

56. The correct answer is B. Ball joints can be pressed, threaded or riveted onto control arms. Riveted joints are usually replaced with bolt on replacements after the rivets are removed.

57. The correct answer is C. Torsion bars are not normally interchangeable from side-to-side. This is because the direction of the twisting or torsion is not the same on the left and right sides.

58. The correct answer is D. If the ballstud has broken, it is possible that the tapered hole in the steering knuckle has become distorted. You can check this by trying the new ball joint stud in the hole: if there is any free-play or if the new tapered stud can rock in the hole, the hole is rounded out. If this has occurred, the steering knuckle must be replaced.

59. The correct answer is A. If the vehicle is equipped with a limited-slip differential, raise the opposite wheel to be safe. Otherwise, the vehicle could either damage the differential or drive itself over the wheel chocks and cause an accident. Technician B is wrong because when wheels have been spin-balanced on the axle, they must be rebalanced whenever the tires are rotated.

60. The correct answer is C, both technicians are correct. Both measurements are commonly used. Consult the vehicle service manual for specifications and measuring locations.

61. The correct answer is A. The wheel speed sensor is used in anti-lock brake (ABS) and traction control systems. All of the other sensors are used in electronically controlled suspension systems.

62. The correct answer is B. During parallel parking maneuvers, the steering wheel is turned against the stops while the vehicle is relatively stationary. This causes high pressure in the power steering system, which places a load on the engine. The power steering pressure sensor informs the PCM of this high pressure, causing the PCM to in turn adjust the IAC valve to raise the engine idle speed. If the sensor is faulty, the PCM will not receive the signal and the engine will stall under the increased load. Technician A is wrong because if the IAC valve was faulty, the vehicle could stall at any time, and not just when parallel parking.

63. The correct answer is A. In an SLA suspension, the coil spring is located between the lower control arm and the vehicle frame. By supporting the lower arm near the ball joint, the weight of the vehicle will keep the coil spring compressed enough to allow removal of the upper control arm. If the lower control arm was being removed, then compressing the coil spring with a spring compressor would be necessary.

64. The correct answer is C, both technicians are right. Faulty brake components can cause a vehicle to pull to one side during braking. Worn strut rod bushings would cause the caster angle to change during braking, also resulting in the vehicle pulling to that side.

65. The correct answer is D. If there is too little worm bearing preload in the steering gear, the steering will feel loose but the toe setting will not be affected. All of the other answers could cause the toe setting to be incorrect.

NOTES

NOTES

Glossary of Terms

--a--

abrasion - wearing away by rubbing or scraping.

Ackerman principle - the geometric principle used to provide toe-out on turns; the ends of the steering arms are angled so that the inside wheel turns more than the outside wheel.

actuator - a device that delivers mechanical energy in response to an electrical signal.

air spring - a suspension device made up of a flexible bladder containing compressed air. The air spring takes the place of a conventional coil or leaf spring. Air is supplied by an on-board compressor, usually with auxiliary equipment to sense vehicle height and modify the pressure in the air spring as needed.

alignment - to bring parts or components into proper coordination.

anti-roll bar - see **stabilizer bar**.

aquaplaning - the loss of traction that results when a tire loses sufficient grip on a wet roadway; the tire actually loses contact with the road surface and rides on a thin film of water.

aspect ratio - the relationship between the height of a tire from bead to tread, and the tread width; usually expressed as a percentage of the tread width.

asymmetric - a lack of correspondence in size or shape between opposed parts.

axial - around, on or along an axis.

axial load - a load applied which is parallel to the axis of a component, such as a shaft.

axial play - movement of a component parallel to its axis.

axis - a real or imaginary straight line on which or around which an object rotates.

--b--

backlash - the play between moving parts.

back pressure - the pressure caused in a system by a blockage or restriction.

ball bearing - a friction-reducing bearing with which the moving parts revolve or slide on freely rolling metal balls; also, any one of the balls.

ball joint - a suspension component that provides a pivot point, allowing the steering knuckle to move up and down as well as turn in response to steering input. The ball fits into a socket housing that is attached to the control arm and the stud on the other end of the ball is attached to the steering knuckle. A dust cover is installed over the ball and socket assembly to keep dirt out and lubricant in.

bead - the steel reinforced inner edge of a tire, which fits inside and seals against the wheel rim.

bearing clearance - the space between a bearing and its corresponding component's loaded surface. bearing clearances are commonly provided to allow lubrication between the parts.

bearing race - the machined surface of a bearing assembly against which the needles, balls or rollers move.

bias - a slanting or diagonal line. In tire construction, referring to the manner in which the belts or lies are laid.

bolt diameter - the measurement across the major area of a bolt's threaded area or shank.

bolt head - the part of a bolt upon which the turning force is applied.

boot - a protective cover, usually rubber, used to shield a component or assembly from contaminants and/or to contain lubricants.

bracket - a supporting device, usually angled, used to secure a component to a structure.

bump steer - a steering problem in which a vehicle tends to the left or the right after a bump, without steering wheel input from the driver. this is usually caused by some steering misalignment or damage that permits change of toe when the suspension works up and down.

bushing - a liner, usually removable, for a bearing; an anti-friction liner used in place of a bearing.

--c--

camber - the attitude of a wheel/tire assembly in which, when viewed from the front, the distance between the tops and bottoms of the tires are different. If the distance between the tops is greater than between the bottoms, positive camber is present; if the distance between the tops is less than between the bottoms, negative camber is present.

capacitor - a device, made up of two or more conducting plates, separated by an insulator, used to store an electric charge.

castellations - slots cut in a bolt head or on nut flanges, through which a cotter pin is inserted to secure the fastener.

caster - the angle formed by the relationship of the kingpin axis and a vertical axis through the wheel centerline when viewed from the side of the vehicle. Positive caster is that which is inclined towards the rear.

center link - a steering linkage component, which attaches the pitman arm to the idler arm, tie-rod or crosslink.

center of gravity - the point in a body or system around which its weight is evenly distributed or balanced.

centrifugal force - the force that tends to pull an object outward when it's rotating rapidly around a center point.

chafing - wearing away by rubbing.

coil spring - spring steel rod wound into a coil that supports the vehicle's weight while allowing suspension movement.

collapsible steering column - a steering column that is designed to collapse, to prevent the column from heavily impacting the driver during an accident.

computer - an electrical device that receives information from sensors and makes decisions based on these inputs along with programmed information, and sends out the decisions to actuators.

concentric - having a common center, such as concentric circles.

control arm - a suspension component that connects the vehicle frame to the steering knuckle or axle housing and allows the up-and-down movement of the wheels.

cords - reinforcing materials running through a tire's plies, usually fiberglass or steel.

corrosion - the eating into or wearing away of a substance gradually by rusting or chemical action.

crossmember - part of the vehicle frame structure, arranged transversely and attached to the frame rails at each side of the vehicle. Can be removable or welded in place.

current - the flow or rate of flow of an electric charge through a conductor or medium between two points having a different potential, expressed in amperes.

--d--

dampen - to lessen oscillation or movement.

data link connector (DLC) - a means through which information about the state of the vehicle control system can be extracted with a scan tool. This information includes actual readouts on each sensor's input circuit and some actuator signals. It also includes any trouble codes stored. The data link connector is also used to disable the computer's ignition timing adjustments on some engines so base or reference timing can be measured with a timing light. Before OBD II, each OEM had a unique data link connector and called it by a different name. With the advent of OBD II, the DLC became standardized as a 16-pin connector to which the scan tool could be connected to read data and sometimes control outputs of the PCM.

dead axle - a load-supporting axle that does not transmit power.

deflection - a turning aside, bending or deviation.

diagnostic trouble code (DTC) - a code that represents and can be used to identify a malfunction in a computer control system.

dial caliper - a measuring device equipped with a readout dial used to measure depth, length or diameter.

dial indicator - a measuring device equipped with a readout dial, used most often to determine end motion or irregularities.

differential - the arrangement of gears connecting two axles in the same line and dividing the force between them, but allowing one axle to turn faster than the other, as when the vehicle turns.

directional stability - the tendency of a vehicle to travel in a straight line on a flat surface without driver control.

Glossary of Terms

directional tires - tires with a tread pattern that is designed to give maximum traction by removing water from under the tread in such a way as to minimize the risk of aquaplaning. Directional tires must be installed to turn in a specific direction.

disable - in automotive terminology, rendering a system inoperative.

DLC - see **data link connector**.

drag link - a steering linkage component that connects the pitman arm and the steering arm.

DTC - see **diagnostic trouble code**.

dynamic balancing - balancing a part while it is in motion.

--e--

eccentric - a rotating part of a shaft that is set off center of the axis.

elasticity - having the property of immediately returning to original size, shape or position after being stretched, squeezed, flexed, expanded, etc.

electromagnet - an iron core surrounded by a coil of wire that temporarily becomes a magnet when an electric current flows through the wire.

electromagnetic induction - moving a conductor through a magnetic field to produce current in the conductor.

electromechanical - a mechanical device or operation that is activated or regulated by electricity.

electronic - pertaining to the control of systems or devices by the use of small electrical signals and various semiconductor devices and circuits.

electronic level control - a suspension system that uses air springs to maintain vehicle ride height. Height sensors are used to signal a control unit when the vehicle is riding low or high. In response to this signal, compressed air is either sent to or vented from the air springs.

end-play - the regulated movement of a component, usually a shaft, during operation.

--f--

foot-pound - a unit of energy required to raise a weight of one pound, a distance of one foot.

free-play - the measurable travel in a mechanical device between the time force is applied and work is accomplished, such as the movement in a linkage assembly.

friction - the resistance to the motion of two moving objects in contact with each other.

fulcrum - the support or point of support on which a lever turns when raising or moving something.

--g--

ground - a connecting body whose electrical potential is zero to which an electrical circuit can be connected.

--h--

hard spot - an area in a casting, which has become harder (more dense) than the surrounding material.

harmonic vibration - periodic motion or vibration along a straight line. The severity depends on the frequency or amplitude.

height sensor - a component used in an air suspension system to signal a control unit when the vehicle is riding low or high. In response to this signal, compressed air is either sent to or vented from the air springs.

hub - the center part of a wheel, gear, etc. that rides on a shaft.

--i--

idle air control (IAC) valve - controls the amount of air allowed to bypass the closed throttle to keep the engine at the proper idle speed. The computer also controls the idle speed of the engine, depending on engine coolant temperature and the number of accessories such as headlights, air conditioning, etc. that are engaged. The idle air control (IAC) valve controls air flow through a throttle bypass passage by means of a stepper motor, an electric motor that can move to a specific location in its travel. Based on the information from its sensors and the parameters in its memory, the computer sends a duty-cycle (percentage of on-time) signal to the IAC valve motor to open or close the bypass to increase or slow the idle speed. This signal can range from zero to 100 percent, and corresponds to the amount of airflow the computer determines is needed. Displayed on the scan tool as counts or percentage.

idler arm - a conventional steering system component consisting of an arm that swivels in a bushing on a shaft, which is attached to the frame. The idler arm is mounted on the right side of the vehicle and is the same length and set at the same angle as the pitman arm. Its function is to hold the right end of the center link level with the left end, which is moved by the pitman arm, and transfer the steering motion to the right side tie-rod.

inch pound - one twelfth of a foot pound.

included angle - the sum of the angle of camber and steering axis inclination; the sum of two intersecting angles.

independent suspension - a suspension in which each wheel can travel up-and-down without directly affecting the position of the opposite wheel.

induction - the process by which an electric or magnetic effect is produced in an electrical conductor or magnetic body, when it is exposed to variation of a field of force. Induction is the principle used in an ignition coil to increase voltage.

integral - made up of parts forming a whole.

integral power steering - a power steering system in which the power cylinder and control valve are contained in one housing.

--k--

kinetic balance - balance of radial forces on a spinning tire; determined by electronic wheel balancer.

kinetic energy - energy in motion; the energy of a body that results from its motion; it's equal to half the product of its mass and the square of its velocity.

kingpin - the pivot shaft for the steering knuckle on most early axles and some modern heavy-duty axles.

knuckle - see *steering knuckle*.

--l--

lateral runout - side-to-side movement or wobble in a wheel or tire.

leaf spring - a suspension spring consisting of a single flat plate made of steel or composite material or several steel plates bundled together.

live axle - an axle on which the wheels are firmly affixed, with the axle driving the wheels.

lubrication - the process of introducing a friction reducing substance between contacting parts to reduce wear.

--m--

Macpherson strut - the principal device in the suspension of the same name, in which the spring and shock absorber are combined in a single unit.

magnet - any substance that attracts iron or an iron, steel or any ferrous metal alloy.

magnetic field - the region of space in which there is a measurable magnetic force.

mechanical advantage - the ratio of the output force of a device performing work, to the input force; used to determine the efficiency of a machine.

memory - part of a computer that stores or holds programs and other data.

memory steer - a steering condition where the steering wheel and wheels want to return to a position other than center. This can be caused by tightening rubber bonded socket tie-rod ends when the steering wheel is not centered, binding in the upper strut mounts, or binding in a steering component or ball joint.

motor - a machine that converts electrical energy into mechanical energy.

multimeter - a measuring device combining the functions of voltmeter, ammeter and ohmmeter for both DC and AC scales.

--o--

offset - positioned off center or at an angle; a curve or bend in a metal bar to permit it to pass an obstruction.

ohm - a unit of electrical resistance of a circuit in which an electromotive force of one volt maintains a current of one ampere, named after German physicist Georg ohm.

ohm's law - a law of physics which states that the steady current through certain electrical circuits is directly proportionate to the applied electromotive force.

Glossary of Terms

on-car balancing - the practice of spinning a wheel on the car to balance the wheel and all other rotational components together.

open circuit - an electrical circuit that has a break, preventing the flow of electrons.

orifice - a precisely sized opening in a tube, cavity, etc.

out-of-round - condition by which an outside or inside diameter has become less than perfectly uniform, usually through wear or temperature extremes.

--p--

parallelogram steering linkage - a type of conventional steering linkage consisting of a pitman arm, center link, idler arm and tie-rod assemblies to connect to the steering knuckles. The pitman arm, center link and idler arm form three sides of a parallelogram.

pitman arm - a steering system component mounted on the steering box shaft that transfers the gearbox motion to the steering linkage.

play - the relative movement between or among parts.

powertrain control module (PCM) - on vehicles with computer control systems, the main computer that determines engine operation based on sensor inputs and by using its actuator outputs. The PCM may also control transmission operation.

preload - calculated force applied to parts, especially bearings, to eliminate play.

pressure - the force exerted against a body, measured in units of force per unit of area.

pull - a steering condition where the vehicle driver has to maintain constant pressure on the steering wheel to keep the vehicle moving straight.

pulse width - the length of time during which a circuit is energized.

pulse width modulated - electronic control of a solenoid that rapidly cycles it on and off many times per second in order to achieve a specific output.

--r--

race - a channel, groove or machined surface for moving parts to bear on, such as a bearing race.

rack-and-pinion steering - a type of steering mechanism that replaces the pitman arm, center link and idler arm on gearbox steering. The steering column ends in a pinion gear that moves the driven rack to the left and right. The rack ends contain ball studs connected to the steering knuckles.

radial - branching out in all directions from a common center.

radial load - the load applied at a right angle to an axis.

radial runout - the out-of-roundness of a wheel or tire.

radius arm - a suspension component that is connected to a twin I-beam or solid axle at one end and to the vehicle frame through bushings at the other. The radius arm braces the I-beam or axle and keeps it at a right angle to the vehicle frame.

ratio - the fixed relation in degrees, number, etc., of two similar things.

rebound - the extension of a spring after compression.

recirculating ball steering - a steering gear in which steering movement is transmitted by a worm gear and ball nut. A set of ball bearings rolls in the grooves between the worm gear and ball nut in a continuous loop.

resistance - in electrical terms, the property of a conductor by which it opposes the flow of an electrical current, resulting in the heating of the conducting material.

ride height - the dimension between a fixed point on the vehicle and the pavement. The fixed point varies according to vehicle and manufacturer. Also called vehicle height.

roller bearing - an anti-friction device made up of hardened inner and outer races between which steel rollers move.

runout - measurable irregularity across a plane surface, such as a disc brake rotor.

--s--

SAI - see **steering axis inclination**.

scalloping - a tire wear pattern (caused by wheel imbalance) in which pieces appear to be cut out of the tire by a spoon.

scan tool - microprocessor designed to communicate with a vehicle's on-board computer system to perform diagnostic and troubleshooting functions.

score - a scratch, ridge or groove on a finished surface; to mar a surface in that way.

scrub radius - the distance between the point at which the tire's vertical centerpoint intersects the road, and the steering axis inclination (SAI) intersects the road.

scuffing - scraping or wearing of a surface in patches.

sensor - any mechanism by which the engine control computer can measure some variable on the engine, such as coolant temperature or engine speed. Each sensor works by sending the computer a signal of some sort, a coded electronic message that corresponds to some point on the range of the variable measured by that sensor.

serpentine belt - a flat, ribbed drive belt that makes multiple angles, driving several components.

shackle - the attachment to the frame for one end of a leaf spring. The shackle allows the spring to change in length as the vehicle encounters uneven road surfaces.

shock absorber - a device used to dampen the oscillation of the suspension caused by irregularities in the road surface.

short and long arm suspension - a suspension system in which the upper control arm is shorter than the lower control arm, allowing the wheel to deflect in a vertical direction with minimal change in camber.

shudder - a momentary shaking or vibration.

spindle - a shaft used to attach the wheel assembly on non-drive axles.

splines - grooves cut into the outside or inside surface of a component to enable it to fit with another component having corresponding grooves.

spring - a suspension system component that supports the vehicle and absorbs shock caused by uneven road surfaces; a device that returns to its original form after being forced out of shape.

sprung weight - the weight of all the vehicle components that are supported by the springs; see **unsprung weight**.

stabilizer bar - a torsion-bar spring connecting the suspension on either side of the vehicle. When a vehicle rolls to the side in a turn, the suspension at the outside wheel compresses and the suspension at the inside wheel extends. The stabilizer bar that connects them twists to apply a counteracting force to hold the vehicle closer to level. Also called an **anti-roll bar** or **sway bar**.

static balance - balance at rest; still balance; the equal distribution of weight of the wheel and tire around the axis of rotation.

steering arm - the steering system component that links the steering knuckle to the tie-rod assembly.

steering axis inclination (SAI) - the angle between true vertical and an imaginary line running through the rotational center of the ball joint(s).

steering column - the housing, steering shaft, bearings and related components between the steering wheel and the steering gear.

steering gear - the assembly located at the end of the steering column, which contains the gears and other components that multiply the driver turning force.

steering knuckle - the suspension component that connects the upper and lower control arms or the strut and lower control arm. On rear-wheel drive vehicles, it usually incorporates the front wheel spindle and on front-wheel drive vehicles it has an opening where the halfshaft passes through. A steering arm is attached to the steering knuckle, where the tie-rod end is connected.

steering linkage - all of the components that connect the steering gear to the front wheels.

strut rod - on vehicles where the lower control arm is attached to the frame at one pivot point, a strut rod is used to brace the control arm against the vehicle frame.

subframe - a removable part of the vehicle frame to which drivetrain, suspension and/or steering components are commonly attached.

sway bar - see **stabilizer bar**.

--t--

taper - a gradual decrease in width or thickness.

tension - stress exerted on an object by a pulling that tends to extend the material.

103

Glossary of Terms

thrust - the continuous pressure of one object against another.

thrust angle - the difference between the thrust line and the geometric centerline of the vehicle.

thrust line - an imaginary line that divides the total to angle of the rear wheels.

thrust line alignment - aligning the front wheels to the thrust line during a wheel alignment, when rear-wheel toe cannot be adjusted to specification.

tie-rod - steering linkage member that connects the steering knuckle with the center link or the steering rack.

tie-rod end - a ball and socket joint that connects the tie-rod to the steering knuckle arm and to the center link or steering rack.

tire rotation - the practice of moving a set of tires to different positions on the vehicle to equalize wear and extend the life of the tires.

toe - the direction in which a wheel tends to roll; a major factor in tire wear.

toe-in - a condition that exists if the tire's line of forward direction intersects the extended centerline of the vehicle.

toe-out - a condition that exists if the tire's line of forward direction and the vehicle centerline are angled apart.

toe-out on turns - the designed angle of the steering arm on the steering knuckle, which causes the inside front wheel to turn at a sharper angle than the outside front wheel during a turn. The specification is checked using the turntables on wheel alignment machine. Toe-out on turns is not an adjustable angle, and if it is incorrect, it is most likely due to a bent steering arm.

tolerance - the difference between the allowable maximum and minimum dimensions of a mechanical part; the basis for determining the accuracy of a fitting.

torque - the twisting effect or moment exerted upon an object by a force operating at a distance from that object; equal to the force multiplied by the perpendicular distance between the line of action of the force and the center of rotation at which it is exerted.

torque steer - the tendency of many front-wheel drive vehicles with half-shafts of unequal length, to turn somewhat from the desired direction when accelerating, especially in a curve, or when decelerating in a curve.

torsion bar - a bar made of spring steel that uses a twisting motion to support the weight of the vehicle and absorb road shock.

track - the distance between the centers of the treads of parallel wheels.

tracking - travel of the rear wheels in a parallel path with the front wheels.

transverse - situated across; crossing from side-to-side; crosswise.

twin I-beam suspension - a type of independent front suspension used on light-trucks and vans. It consists of two I-beams supported by coil springs, and the steering knuckles/spindles, which are connected by kingpins or ball joints. The inner end of the axle connects to the vehicle frame through a rubber bushing. A radius arm also connects to the frame through rubber bushings to control wheelbase and caster.

--u--

unsprung weight - the components of a vehicle which rest directly on the road surface without being supported by the suspension springs.

--v--

vehicle height - see *ride height*.

vehicle speed sensor (VSS) - a permanent magnet sensor, usually located on the transmission, that provides an input to the vehicle computer control system regarding vehicle speed.

--w--

wheel alignment - the adjustment of suspension and steering components to optimize steering control and minimize tire wear.

wheel balance - the condition in which a wheel/tire assembly has equal weight around its center, preventing vibration at high speeds. Wheel balance can be static, such as on a bubble balancer, or dynamic, such as with a spin balancer.

wheel offset - the dimensional difference between a wheel's centerline and the plane of the axle flange mounting surface.

wheel weights - small weights, usually made of lead, attached either mechanically or by adhesive to a wheel/tire assembly to correct its balance.

worm gear - a gear into which teeth are cut, resembling the threads of a screw.

NOTES